AI for Radiology

Artificial intelligence (AI) has revolutionized many areas of medicine and is increasingly being embraced. This book focuses on the integral role of AI in radiology, shedding light on how this technology can enhance patient care and streamline professional workflows.

This book reviews, explains, and contextualizes some of the most current, practical, and relevant developments in artificial intelligence and deep learning in radiology and medical image analysis. *AI for Radiology* presents a balanced viewpoint of the impact of AI in these fields, underscoring that AI technologies are not intended to replace radiologists but rather to augment their capabilities, freeing professionals to focus on more complex cases. This book guides readers from the basic principles of AI to their practical applications in radiology, moving from the role of data in AI to the ethical and regulatory considerations of using AI in radiology and concluding with a selection of resources for further exploration.

This book has been crafted with a diverse readership in mind. It is a valuable asset for medical professionals eager to stay up to date with AI developments, computer scientists curious about AI's clinical applications, and anyone interested in the intersection of healthcare and technology.

Oge Marques, PhD, is Professor of Computer Science and Engineering in the College of Engineering and Computer

Science, Professor of Biomedical Science (*Secondary*) in the Charles E. Schmidt College of Medicine, and Professor of Information Technology (*by courtesy*), in the College of Business at Florida Atlantic University (Boca Raton, FL – USA).

He is the author of 12 technical books, one patent, and more than 130 refereed scientific articles on image processing, medical image analysis, computer vision, artificial intelligence, and machine learning. He is a senior member of both the Institute of Electrical and Electronics Engineers and the Association for Computing Machinery, Fellow of the National Institutes of Health AIM-AHEAD Consortium, Fellow of the Leshner Leadership Institute of the American Association for the Advancement of Science, Tau Beta Pi Eminent Engineer, and member of the honor societies of Sigma Xi, Phi Kappa Phi, and Upsilon Pi Epsilon.

AI for Everything

Artificial intelligence (AI) is all around us. From driverless cars to game winning computers to fraud protection, AI is already involved in many aspects of life, and its impact will only continue to grow in future. Many of the world's most valuable companies are investing heavily in AI research and development, and not a day goes by without news of cutting-edge breakthroughs in AI and robotics.

The *AI for Everything* series explores the role of AI in contemporary life, from cars and aircraft to medicine, education, fashion and beyond. Concise and accessible, each book is written by an expert in the field and will bring the study and reality of AI to a broad readership including interested professionals, students, researchers, and lay readers.

For more information about this series please visit: https://
www.routledge.com/AI-for-Everything/book-series/AIFE

AI for Radiology

Oge Marques, PhD

CRC Press
Taylor & Francis Group
Boca Raton London New York

CRC Press is an imprint of the
Taylor & Francis Group, an **informa** business

First edition published 2024
by CRC Press
2385 NW Executive Center Drive, Suite 320, Boca Raton FL 33431

and by CRC Press
4 Park Square, Milton Park, Abingdon, Oxon, OX14 4RN

CRC Press is an imprint of Taylor & Francis Group, LLC

ISBN: 978-0-367-62778-2 (hbk)
ISBN: 978-0-367-62725-6 (pbk)
ISBN: 978-1-003-11076-7 (ebk)

DOI: 10.1201/9781003110767

Typeset in Palatino

by Apex CoVantage, LLC

To my beloved Ingrid, with deep gratitude for her unconditional

love, unlimited patience, and constant encouragement

Contents

Contents

Foreword

In the realm of radiology, a nexus has emerged between advanced machine learning methodologies and the intricate art of medical imaging. It's here, at this confluence, that I had the pleasure of meeting a luminary in the field, Dr. Oge Marques. Our paths first crossed at a SIIM (Society for Imaging Informatics in Medicine) meeting, an occasion that would mark the beginning of numerous fruitful collaborations.

As a neuroradiologist deeply entrenched in the applications of AI in medicine, my journey has allowed me to interface with various tools and innovations. But collaborating with Oge has been an education unto itself. With his robust background in computer science and a prolific portfolio of publications, Oge doesn't just utilize AI; he advances it. His approach, underpinned by rigorous scientific methodology, is a testament to his commitment to both the precision of computer science and the compassion intrinsic to healthcare.

The content of this book, judiciously curated by Oge, is both comprehensive and enlightening. Beginning with a broad overview of AI, especially its increasing importance in healthcare, it delves into the specific applications and opportunities of AI in radiology. The fundamentals of machine learning and deep learning are laid out with clarity, making it accessible to both the novice and the seasoned practitioner.

What I find particularly commendable is Oge's insight into medical image analysis. The chapters are a journey, introducing readers to various imaging modalities and guiding them through the intricate process of medical image analysis using advanced AI techniques. But beyond the technicalities, he acknowledges the core of radiology – the data. With a dedicated segment on data – its collection, annotation, preprocessing, and the ethical considerations surrounding its use – Oge accentuates its centrality in the world of AI in radiology.

This book is not just about the present state of affairs. It offers a vision, exploring the future trajectories of AI in radiology, addressing challenges, controversies, and the endless possibilities on the horizon.

Having witnessed Oge's dedication and forward-thinking approach firsthand, I am confident that this book will serve as an invaluable resource. For those stepping into the realm of AI in radiology or seeking to deepen their knowledge, this book provides a holistic, scientifically rigorous, and practical guide.

In an era of digital transformation, with large health organizations such as mine keenly looking towards applied innovation and AI, Oge's contributions become ever more relevant. His ability to anticipate and address pressing challenges has not only benefited our field but also shaped its future direction.

It is both an honor and pleasure to introduce this book to you. I wholeheartedly believe that it will stand as a cornerstone for all enthusiasts eager to delve into the world of AI in Radiology.

Felipe Kitamura, MD, PhD
Director of Applied Innovation and AI at Dasa
Affiliated Professor of Radiology at
Universidade Federal de São Paulo
São Paulo, Brazil, July 2023

Preface

We are currently witnessing an unprecedented surge in the application of artificial intelligence (AI) within the medical field, stretching across a broad range of specializations and poised to reshape various facets of healthcare delivery. As we look forward, we envision AI-based systems evolving to glean optimal therapeutic pathways directly from electronic health records and other sources of data, ushering us into a new era of individualized care and inching closer to the realization of *P4 Medicine* – an approach to medicine that's Predictive, Preventive, Personalized, and Participatory. AI's ability to derive insights from vast datasets will enable us to predict disease risk and progression, guide prevention strategies, customize treatment plans, and foster patient involvement in their healthcare journey. The dawn of AI in medicine is not just about advanced technology; it's about ushering in a new era of healthcare that is truly patient centric and data driven.

For now, certain medical disciplines are poised to experience the immediate influence of these novel developments, predominantly those that rely heavily on visual data interpretation for predictions, diagnosis, and treatment decisions. These include, but are not limited to, pathology, dermatology, ophthalmology, and radiology – the focus of this book.

Radiology is a particularly promising field for AI, as it is a discipline that is already heavily reliant on image analysis. AI can be used to automate many of the tasks that radiologists currently perform, such as image interpretation, diagnosis, and reporting. This can free up radiologists to focus on more complex cases and provide better patient care.

The radiology community has taken quick and effective action to embrace AI, moving from potential fear and alarmism to a position where the consensus within the radiology community is that AI will not replace radiologists; instead, it will augment radiologists' abilities and assist them in some of their tasks. AI applications in radiology can cover a vast spectrum, from the decision to request an imaging study for a patient based on the patient's health record to the reporting and communication of results that might inform further actions.

AI advancements in radiology have the potential to bring considerable improvements to patient care, radiologist workflows, and physician referrals. However, this transformative path is paved with both high expectations and substantial hype. The technical intricacies of AI tools only represent a portion of the equation. For a complete adoption, these tools must be demonstrably effective, reliable, and safe in a clinical environment. They should also uphold the highest ethical standards and comply with regulatory norms to earn full trust.

Implementing AI into radiology practices demands a thorough understanding and careful consideration of the technical, legal, and ethical implications. While AI can provide valuable assistance to radiologists, it doesn't serve to replace human expertise. Instead, it amplifies it. I foresee a future where the radiologist takes on a role analogous to a symphony conductor, steering AI solutions – the orchestra members – each contributing their unique input under the guided coordination of the radiologist's expertise.

This book reviews, explains, and contextualizes some of the most current, practical, and relevant developments in AI and deep learning (DL) in radiology and medical image analysis.

This book is structured as follows.

Chapter 1 introduces and motivates the use of AI and DL in the medical domain and discusses how successful applications of AI in healthcare require a different recipe than the one often used to achieve success in AI in general.

Chapter 2 provides concrete examples of how the radiology community has taken quick and effective action to embrace the new era of AI.

Chapter 3 covers the fundamental concepts and terminology associated with machine learning and DL, the predominant paradigms of contemporary AI.

Chapter 4 provides an overview of the field of radiology from the viewpoint of the medical image analysis pipeline, including fundamental concepts and terminology and main imaging modalities, and offers perspective on the impact of AI and DL in this ever-growing research field.

Chapter 5 is devoted to the essential role of data in AI solutions for radiology. It covers various aspects of data collection, types and sources of data, preprocessing and annotation techniques, data challenges, considerations, and ethical considerations related to data sharing and collaborations.

Chapter 6 shows examples of clinical applications of AI in radiology – both interpretative and noninterpretative – with a focus on commercially available solutions approved for the European and/or American market.

In Chapter 7, Dr. Michèle Retrouvey and I discuss the challenges and initiatives associated with integrating AI content into the medical education curriculum, at undergraduate and graduate levels, with emphasis on radiology.

For the most technically inclined, Chapter 8 offers a learning path to starting with DL in medical imaging. It covers some of the most popular tools, languages, and frameworks for developing DL solutions for medical image analysis today. It also describes an example of a typical DL workflow in radiology.

Chapter 9 takes us into the future of AI in radiology with a focus on three main aspects: (i) ethical, cultural, and regulatory aspects of the use of AI in radiology; (ii) challenges, controversies, and objections to the use of AI and DL models in radiology; and (iii) emerging trends, challenges, and opportunities.

Finally, in Chapter 10, I present a list of curated resources (books, online courses, scientific articles, software tools, and more) to go deeper into some aspects discussed in this book and assist you in the lifelong learning journey that you are about to embark on.

I hope that this book will serve as a quick, sound, and objective scientific reference for readers from both sides of the equation: medical professionals who want to stay current with the latest developments of AI and related technologies, as well as computer scientists and medical imaging informatics professionals who want to learn more about the clinical applications and implications of the use of AI in radiological practice and integration into its workflow.

How to Read This Book

This book has been written with a diverse readership in mind. Therefore, your journey through the chapters may differ based on your specific background and objectives.

For those unfamiliar with the landscape of AI in radiology, the first two chapters will serve as a helpful primer.

Following that, based on your expertise, you may consider bypassing Chapter 3 (if you already have a strong understanding of AI) and/or Chapter 4 (if you are well versed in radiology and medical image analysis).

Chapters 5 through 9 can be read independently and in any order that aligns with your priorities.

Last, Chapter 10 is highly recommended for all readers, and it would be beneficial to earmark some of the resources compiled there for future reference.

Acknowledgments

I was trained as an engineer but have been engaged in projects, grants, and publications related to biomedical sciences since my early years in engineering school. I have been working with AI in some form or another for 35 years.

The impetus for this book came, curiously enough, from my work in dermatology. In 2017, I attended my first *Society for Imaging Informatics in Medicine* (SIIM) Annual Meeting to present research results in skin lesion detection, segmentation, and classification. During that event, I had the privilege of attending numerous keynotes, technical sessions, and hands-on labs in radiology and was welcomed by this amazingly friendly and inclusive community of medical imaging informatics experts. From that moment on, I intensified my efforts in *Medical AI*, became a SIIM member – as well as a member of the *European Society of Medical Imaging Informatics* (EuSoMII), *European Society of Radiology* (ESR), and *Radiological Society of North America* (RSNA) – and along the way had the immense privilege of working closely with some of the brightest minds in this space.

I am immensely grateful to many colleagues in the AI, radiology, and medical imaging informatics communities for their encouragement and valuable lessons throughout these years.

Many thanks to Luciano M. Prevedello, Kevin Mader, Barbaros Selnur Erdal, Igor R. dos Santos, Ian Pan, Felipe C. Kitamura, Bradley J. Erickson, Katherine P. Andriole, George L. Shih, and Peter Chang for their excellent hands-on sessions and labs during several SIIM and RSNA meetings.

I am grateful to Sameer Antani and his group at the National Library of Medicine (NLM) at the National Institutes of Health (NIH), where I had the privilege of spending a sabbatical semester in 2018.

A very special thank you to my dear colleague Les Folio, whom I first met in 2018 and have since then collaborated in several publications and projects, including a very successful SIIM Hackathon in 2019.

Many thanks to my European colleagues, particularly Erik Ranschaert, Peter van Ooijen, Daniel Pinto dos Santos, Elmar Kotter, Angel Alberich-Bayarri, Bettina Baessler, Sergey Morozov, Benoit Rizk, Luis Martí-Bonmatí, and Federica Zanca, for the numerous discussions and exchanges of ideas over the past few years and for welcoming me into the EuSoMII community.

My gratitude to colleagues in the Florida Atlantic University Charles E. Schmidt College of Medicine, where I have had the honor of serving as Professor of Biomedical Science (Secondary), particularly Janet Robishaw, Julie Pilitsis, Adam Wyatt, and Phillip Boiselle.

I am very grateful to Matt Lungren for opening the door to successful collaborations with Stanford University in publications, research, and educational initiatives.

I am also indebted to Pranav Rajpurkar, Tessa Cook, Michael Do, Raym Geis, Judy Wawira Gichoya, Hugh Harvey, and Stephen Borstelmann. I have learned a lot from you and appreciate your willingness to explore exciting paths of scientific discovery.

Special thanks to my colleague Michèle Retrouvey, board-certified radiologist and Director of Radiology Education at the Charles E. Schmidt College of Medicine at Florida Atlantic University in Boca Raton, Florida, for agreeing to co-author Chapter 7 and sharing her insights on why and how AI content should be taught to future doctors and radiologists.

I would like to express my gratitude to Christian Garbin and Mario Taschwer for their comments and suggestions during the preparation of this manuscript and to Humyra Chowdhury, Roshni Merugu, and Jennifer Gogova for their assistance with diagrams, figures, and supporting tasks.

A very special note of gratitude to Randi (Cohen) Slack and her team at CRC Press / Taylor & Francis for their support throughout this project.

Oge Marques
Boca Raton, Florida
July 2023

1

Artificial Intelligence and Medicine: The Big Picture

1.1 Introduction

Artificial intelligence (AI) has been one of the most significant technological developments of our era. In the last ten years, thanks to the emergence of *deep learning* (DL),[1] many high-profile problems that seemed beyond the reach of computer-based solutions have been solved, including better-than-human performance in games such as chess, Go, and Shogi [1]. In parallel with headline-grabbing advancements, AI has become a fundamental tool in the business world, where managers are busy figuring out how to "add AI" to their products and processes to get an advantage over their competitors. AI has impacted every sector of the economy and every field of human activity, from education to transportation, games to communications, shopping to healthcare.

The growing adoption of AI products and solutions in healthcare will affect patients, insurance providers, drug companies, researchers, and medical professionals in every area of medicine. It is expected that in the near future, every medical student will have a significant amount of AI-related content in their curriculum [2], preparing

---

[1] Deep learning (DL) is a term used to refer to AI solutions that use neural networks and are capable of learning from (vast amounts of) data. Since the early 2010s, most AI advancements have been possible thanks to the use of DL approaches. See Chapter 3 for more details.

DOI: 10.1201/9781003110767-1

themselves for a world in which doctors who understand AI will be more valuable than doctors who don't. This is particularly true for radiology, a specialty that, for a brief while, might have felt threatened by the advancements in DL AI and eventually became the driver of progress in the field (see Chapter 2).

Ideally, the future of healthcare will be built by combining the talent of data scientists, machine-learning engineers, statisticians, software developers, and AI researchers on one end with the subject matter expertise of every specialty in medicine on the other end. We should be prepared to build a pathway that will ultimately improve patients' (and doctors') well-being [3].

In this chapter, we introduce and motivate the use of AI in the medical domain and discuss how successful applications of AI in healthcare require a different equation than the one often used to achieve success in AI in general. We start by providing an overview of AI to better understand its foundations, revisit some of its historical ups and downs, standardize the terminology, and appreciate what contemporary AI solutions can and cannot do.

1.2 AI: Foundations, Brief History, and Current State of the Art

1.2.1 Defining AI

Defining AI is surprisingly difficult, in significant part because we don't have a good definition for *intelligence* to begin with, and the use of the word *artificial* in this context is not always precisely clear either. In general, *artificial* usually means "by means of a computer or robot," which suggests that some hardware and software components enable an electronic device to act as if it had cognitive capabilities comparable to the ones exhibited by humans.

A dictionary (*Merriam-Webster*) defines AI as "a branch of computer science dealing with the simulation of intelligent behavior in computers" or "the capability of a machine to imitate intelligent human behavior." An encyclopedia (*Encyclopedia Britannica*) may phrase it as "the ability of a digital computer or computer-controlled robot to perform tasks commonly associated with intelligent beings." These are fine definitions that grab the general sense of the term *AI*.

In more technical texts [4], there are four *groups* of definitions, reflecting some of the most prominent approaches to AI in its history. We follow these book authors in choosing the "rational agents" approach, which sets the goals of AI to be "the study and construction of agents that *do the right thing*" (emphasis added) [4].

1.2.2 Agents and Environments

According to Russell and Norvig,

> [A]n agent is anything that can be viewed as perceiving its environment through sensors and acting upon that environment through actuators.[2] For each possible percept sequence, a rational agent should: select an action that is expected to maximize its performance measure, based on the evidence provided by the percept sequence and whatever built-in knowledge the agent has. [4]

Figure 1.1 shows how rational agents interact with their environments. The box with the question mark is where "the magic happens": it represents the intelligent algorithms capable of processing sensor data (raw input) and producing decisions that will eventually result in actions

---

[2] A *sensor* is a device that detects and measures physical or environmental properties, converting them into electrical signals or other forms of output. An *actuator* is a component that receives a control signal and initiates a physical action or movement, typically to manipulate or control a system or its environment.

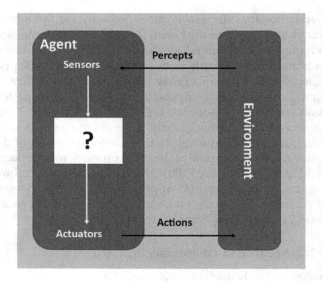

Figure 1.1
Agents and environment (adapted and redrawn from [4]).

in the environment. The diagram is general enough to be used for significantly different scenarios, from the simple game of tic-tac-toe (where the *environment* is fully observable, static, and discrete; the *input* is the current state of the board; the *action* is playing an "X" or "O" in an empty slot; and the *algorithm* can be as simple as a game tree), to the complex world of self-driving vehicles (where the *environment* is partially observable, dynamic, and continuous; the *input* consists of a representation of a rich and vibrant three-dimensional (3D) world of cars, pedestrians, traffic signs, lane marks, and much more, through multiple sensors that produce vast amounts of raw data in different modalities – such as radar, lidar, camera pixels; the *action* might include braking, turning, etc.; and the *algorithm* must be incredibly complex, requiring multiple sub-blocks to handle specialized tasks – from 3D reconstruction of the world to path planning).

In the context of healthcare, the diagram in Figure 1.1 can also be helpful to understand the recently proposed model for a "virtual medical coach" that takes as a percept a vast amount of data associated with a patient – from genomics to medical history, from physical activity and sleep indicators captured by smartwatches or equivalent devices to advances reported in the medical literature that might be relevant to the case at hand – feeds this into a DL AI solution, and produces personalized "virtual health guidance" at the output [5].

1.2.3 AI and Deep Learning

Before we delve deeper into DL, it seems appropriate to set the stage by briefly introducing DL within the broader framework of modern AI. The ongoing AI revolution is being propelled by vast amounts of data and increasing computational power, with DL playing a central role in this dynamic environment.

In the simplest terms, DL is a technique that trains computers to mimic the human brain's decision-making process. DL is a subset of machine learning, a branch of AI that involves training algorithms, or set instructions, to learn from data and make predictions or decisions without being explicitly programmed. DL takes this concept even further by creating deep neural networks – multilayered "artificial brains" – that learn to identify patterns and interpret data by training on vast information.

Imagine a deep neural network as an incredibly complex web of interconnected points, similar to the neurons in our brain. Data inputs travel through this network and are processed at multiple layers, each learning to recognize increasingly complex features. Eventually, the network generates an output, which could be anything from a recommendation for a product to a diagnosis of a medical condition.

In the modern context, AI has evolved into gathering and harnessing an extensive dataset to train and fine-tune these deep neural networks. The goal is to enable these networks to make predictions, solve problems, and learn independently.

Chapter 3 expands upon these concepts and shows how they translate into creating DL solutions.

1.2.4 Promises, Predictions, and Winters

The history of AI can be traced back to a seminal event in the summer of 1956 at Dartmouth College. During this summer workshop, luminaries such as John McCarthy, Marvin Minsky, Nathaniel Rochester, Claude Shannon, and colleagues laid the groundwork for AI as an academic field and explored the potential of computational systems to exhibit intelligent behavior.

Over the next 50 years, the field of AI underwent two complete cycles characterized by periods of enthusiasm, optimism, and ambitious assertions, followed by disillusionment and unmet expectations. These downturns in the cycles are commonly referred to as "AI winters." The general interest in AI, including publications, funding levels, and more, experienced significant declines during these intervals. To illustrate the extent of the decline, the mere mention of terms like *AI* or *neural networks* could be viewed unfavorably in the context of a scientific paper.

The first AI winter occurred from the mid-1970s to the early 1980s [6]. This period was marked by a disconnect between the computational demands of certain theoretical AI solutions and the limitations of the available hardware, software, and network infrastructure required for their implementation. The second AI winter, which took place from the late 1980s to the early 1990s [6], was characterized by the increasing cost, sluggish performance, and difficulty in updating expert system computers compared to the more advanced desktop computers of that time. As a result, funding and resources were redirected toward more promising projects.

1.2.5 AI Today

Since the early 2010s, AI has experienced a significant resurgence, mainly due to several critical factors described next.

One of the primary drivers has been the advances in computer hardware. The development of powerful architectures specifically designed for processing high-dimensional data has been instrumental. Graphics processing units (GPUs), tensor processing units (TPUs), and data processing units (DPUs) have significantly accelerated AI computations, enabling more efficient training and inference processes.

Another crucial factor has been the availability of massive volumes of data. The internet has provided a wealth of information through public pages and social media, and the increasing adoption of electronic records across various businesses, including healthcare, has further contributed to this data pool. This wealth of data has provided the necessary input for AI systems to learn and improve.

The field has also benefited from continuous advancements in algorithms. The development of novel neural network architectures and the optimization of existing algorithms used for training these networks have been vital in enhancing the performance and capabilities of AI systems.

The publicity generated by solving highly visible and challenging problems using machine learning approaches has also played a role in AI's resurgence. Notable examples include AlphaGo and its successor AlphaZero, which demonstrated exceptional performance in the games of Go, chess, and Shogi. These achievements have highlighted the potential of machine learning techniques in solving complex problems.

Finally, the proliferation of better, smaller, cheaper, and widely distributed sensors, often referred to as the *Internet of Things* (IoT), has been a significant factor. These sensors, ranging from smartwatches to cameras, generate vast amounts of data and enable the collection of real-time information. This has facilitated various AI applications in healthcare, transportation, and environmental monitoring.

1.2.5.1 Comparing AI Solutions

The hype surrounding AI, particularly in mass media, often obscures the actual capabilities and limitations of AI systems. Recognizing that *not all AI systems are created equal* is crucial. When assessing the claims made about an AI solution, we suggest considering the following five dimensions along with their associated questions [7]:

1. *Strength:* How intelligent is it?
2. *Breadth:* Does it solve a narrowly defined problem, or is it general?
3. *Training:* How does it learn?
4. *Capabilities:* What kinds of problems are we asking it to solve?
5. *Autonomy:* Are these assistive technologies, or do they act on their own?

Using AlphaGo and AlphaZero as examples, this is how the previous questions could be answered:

1. For games such as chess, Go, and Shogi, it is possible to benchmark AI solutions against one another (as well as against the human-level performance of chess grandmasters, for example). This is done

using the Elo metric.[3] According to that metric, AlphaZero is superior to every other human or AI player in all of these games [8].

2. Both AlphaGo and AlphaZero solve narrowly defined problems: the former learned how to play the game of Go, whereas the latter is slightly less narrow since it also learned how to play chess and Shogi. While AlphaZero demonstrates exceptional performance across different games and exhibits impressive learning capabilities, it cannot be considered as achieving artificial general intelligence (AGI) due to its domain-specific nature. AGI refers to a system with human-like intelligence that can excel in a wide range of tasks, including those outside the scope of specific domains, which AlphaZero does not encompass.

3. Both AlphaGo and AlphaZero use an approach known as "deep reinforcement learning." The main difference between them, however, is that while AlphaGo learned from a dataset of games of Go played by human experts, AlphaZero essentially learned from scratch (hence the *Zero* in its name) by playing millions of games (of Go, chess, or Shogi) against itself and figuring out which strategies worked out best. This has led to AlphaZero being capable of choosing movements that were somewhat unusual, surprisingly effective, and downright puzzling to human experts in those games.

4. AlphaZero is designed to solve complex problems related to strategy, decision-making, and optimal play in board games. Specifically, it aims to tackle

---

[3] Elo metric is a rating system used to evaluate the relative skill levels of players in competitive games, initially designed for chess but now applied to various other sports and games as well.

challenges in games such as Go, chess, and Shogi, where the primary objective is to make intelligent moves, outmaneuver opponents, and ultimately win the game. AlphaZero focuses on mastering the intricacies and strategies involved in these specific games rather than addressing broader problem domains outside the realm of board game playing.

5. AlphaGo and AlphaZero are truly autonomous technologies; the only "assistance" they might need is to place physical pieces on the respective physical board.

To conclude this brief discussion about the five dimensions of AI solutions, it seems useful to mention that

- Not all AI solutions can be benchmarked against each other – or compared to humans performing the same task. This is why open challenges and competitions such as ImageNet LSVRC, Kaggle challenges, and radiology-specific challenges (see Chapter 2) have become essential factors in driving the evolution of AI in recent years. In other words, for most AI applications, it is impossible to rank solutions based on their "strength"; the universe of chess, go, and Shogi is a notable exception.

- Most AI solutions available today, however impressive, are examples of "narrow AI" (although some are narrower than others, as the AlphaGo/AlphaZero example demonstrated). This is quickly changing, though, since late 2022, with the popularization of advanced large language models (LLMs), such as OpenAI's GPT-3.5 and GPT-4 (the models behind the – now ubiquitous – ChatGPT). The search for the elusive goal of artificial general intelligence (AGI) (also referred to as "strong AI") continues, and the debate on how soon we might be reaching that stage is heating again.[4]

- Discussions about how an AI solution learns before a model is built and deployed (e.g., how much data are needed, where the data come from, how can we ensure that the data are representative of the population, etc.) as well as its ability to continue to learn and improve its performance after it has been deployed, are fundamentally important. We revisit some of these issues in this context of radiology later in this book.

- When developing AI solutions, it is essential to start by asking meaningful questions that will promote a clear understanding of which problems we expect to solve, leading to better user and system specifications, a smoother software development process, better documentation, and improved workflow. Failure to do so, on the other hand, may lead to inaccurate evaluation, misleading claims of success, and overall distrust.

- For most applications, especially in radiology/medicine/healthcare, it is common (and convenient) to think of AI as "augmented intelligence," removing the cumbersome word "artificial" from the equation and suggesting a collaborative ecosystem where AI and humans work together toward common goals – much like medicine itself, where multiple professionals apply their collective expertise for the benefit of the patient.

1.2.5.2 What AI Can (and Cannot) Do Well

The enormous hype around AI in the media – both when its successes are artificially magnified as well as when the

---

[4] See this "AI Apocalypse Scorecard" recently compiled by *IEEE Spectrum* magazine [9] for a quick overview.

prospects of job losses and other societal impacts due to AI are exaggerated – makes it difficult to draw a precise line between what contemporary AI solutions can do well and what they cannot. The lists that follow (adapted from [10]) should help.

Today's AI *can:*

- Perform some (often impressive) well-defined tasks as well as or better than humans. Examples include image classification, speech recognition, handwriting transcription, and autonomous driving.
- Find and act upon patterns in data, including patterns invisible to humans.
- Get better at performing certain tasks when given lots of labeled, well-organized data from which to learn.

Today's AI however typically *cannot:*

- Perform any entire job better than humans can.
- Explain its mechanism for finding patterns in information or what those patterns mean.
- Understand the context that surrounds a given task.
- Perform tasks that require creativity, empathy, or complex judgment.

In summary, looking at what today's AI can and cannot do, we start to realize that, as AI technology advances and some of its limitations become apparent, it would be advisable to rethink the meaning of the "A" in "AI": rather than *artificial* intelligence (and the associated negative implications in many people's minds), today's AI is being used to assist, augment, and automate processes in a wide variety of areas. Perhaps what we have today – and will continue to see more of in the near future – is akin to "automated intelligence," "augmented intelligence," or "assistive intelligence," all of which are excellent examples of putting technology to good use and (hopefully) improving human lives and optimizing the use of exhaustible resources along the way.

1.3 AI in Healthcare

1.3.1 AI in Healthcare: The Growing Potential

There has been explosive growth in the field of AI in healthcare in recent years, whether it is measured by the number of scientific publications, mentions in the mass media, filed patents, or venture capital (VC) deals (and dollar amounts) for funding start-up companies in this space. The editorial of the 25th anniversary edition of the prestigious *Nature Medicine* journal whose central theme was "Medicine in the Digital Age" states that "Digital medicine . . . holds promise in revolutionizing healthcare and well-being. . . . AI, and in particular deep learning (DL), is among the leading technological tools beginning to be used in the interpretation of medical images and electronic health records" [11].

In short, *AI in healthcare is here to stay*. There are numerous references and vast amounts of data to support this claim. They range from scientific discoveries (publications, grants, citations, patents, etc.) to entrepreneurial activity (start-up companies and associated VC amounts and the number of deals, etc.) to deployment, clinical validation, and US Food and Drug Administration (FDA) clearance of AI solutions.

It should be noted that *AI in healthcare* has multiple facets and dimensions, many of which are beyond the scope of this book (e.g., surgical robotics, clinical trial participation, or fraud detection, to name but a few). The adoption of AI and DL solutions will vary among different medical specialties and will impact different aspects of the healthcare system at different stages, and the prevalence of AI methods will span human life, from embryo selection for in vitro fertilization [12] to prediction of 24-hour mortality in palliative care [13].

Eventually, smart healthcare systems will emerge, in which "the best treatment decisions are computationally learned from electronic health record data by deep-learning methodologies" [14] Until then, certain areas of medicine are more likely than others to be impacted by the latest developments in DL AI, particularly those areas in which predictions, prognoses, and decisions are made based primarily on visual data, notably pathology, derma-tology, ophthalmology, and radiology.

1.3.2 AI in Healthcare: A New Formula

The recipe for success in AI (in general) includes ingredients such as the availability of large amounts of good quality data, access to powerful computational resources, devel-opment and validation of robust algorithms, and massive investments to make it all come together. We can think of it as an equation[5]:

$$\text{Success in AI} = \text{Big Data} + \text{Powerful Compute} + \text{Robust Algorithms} + \text{Massive Investments}$$

These four ingredients are necessary, but not sufficient, for achieving successful results in AI in healthcare. AI in healthcare requires a new equation, where the AI efforts and investments are aligned with the goals, vision, knowledge,

---

[5] This equation, and the ones that follow it, were inspired by a slide by Dr. Keith Dreyer presented during his keynote at the *Society for Imaging* Informatics in Medicine 2017 Annual Meeting, Pittsburgh, PA.

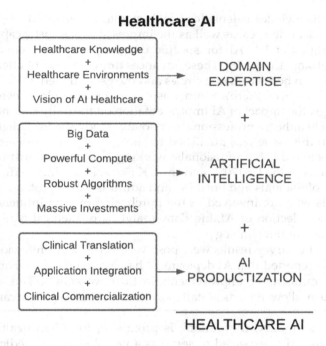

Figure 1.2
AI in healthcare requires a new equation (adapted and redrawn from a slide used by Dr. Keith Dryer in his keynote speech at SIIM 2017 Annual Meeting, Pittsburgh, PA).

and environments for which those AI solutions might be useful ("Domain Expertise" in Figure 1.2) and eventually result in products that successfully meet the requirements for clinical translation, workflow integration, and eventual commercialization ("AI Productization" in Figure 1.2).

1.3.3 AI in Healthcare: Encouraging Early Results for Medical Professionals

In the discourse surrounding AI's role in healthcare, the focus primarily lies on new technological breakthroughs.

This includes patents for health-monitoring methods, sensors, or devices, as well as the impressive diagnostic capabilities of DL AI for specific diseases, such as diabetic retinopathy [15,16]. These advances draw significant attention in both academic circles and the general media.

However, there's another aspect that receives less coverage: the impact of AI implementation on the daily routines of healthcare professionals. A notable study shedding light on this issue was published in late 2019 [17]. The research included 908 professionals[6] working at healthcare institutions in the US (70%) and the UK (30%), including medical professionals and business and administrative professionals who are involved in the purchasing or who influence the selection of AI, big data analytics, or medical equipment and technology.

The survey results were positive. Over 82% of professionals reported that AI deployment had led to operational and administrative improvements in their workflows. This, in turn, allowed medical staff more time to focus on patient care.

In summary, the outlook is promising for AI in healthcare. AI is expected to serve as a valuable aid in medical procedures while always necessitating human supervision and oversight. There is no need to harbor fear toward AI; instead, we should focus on being prepared. The Radiology community is an outstanding example of how this preparedness can be accomplished, as we explore in Chapter 2.

---

[6] Of the total, 17% are medical doctors and specialists, 5% are nurses or nurse practitioners, 26% are senior management, 16% are in information technology, 16% are in research and development, 9% are in legal or regulatory departments, and 9% are in finance or accounting.

Key Takeaways

- Despite the hype, AI is poised to impact numerous aspects of healthcare and several medical specializations, notably radiology.

- AI in healthcare will likely play an assistive role in medical routine and will always require human supervision and oversight.

- AI has the potential to impact every aspect of the healthcare system.

- This is a highly active research area, but the application of research results to the real world is just beginning.

- Success in AI in healthcare requires a different equation than the one often used to achieve success in AI in general.

2

AI in Radiology: From Fear to Leadership

2.1 Introduction

In 2016, at the *Machine Learning and the Market for Intelligence Conference* in Toronto, the prominent researcher Geoffrey Hinton[1] declared that "We should stop training radiologists now. It's quite obvious that in five years deep learning will do better than radiologists" [18]. He was speaking about the potential of deep learning (DL) to revolutionize radiology, and he argued that radiologists were at risk of being replaced by artificial intelligence (AI) systems within five years. Hinton's statement was based on the success of DL algorithms in other medical imaging tasks, such as classifying skin cancer and detecting diabetic retinopathy. He argued that these algorithms were already better than radiologists at identifying specific types of diseases and that they would only get better with time.

This bombastic statement came at a time when other news outlets were also speculating about the rise of "robot radiologists" [19], and even scholarly journals in the field contemplated the threat of AI, machine learning (ML), and DL to the radiology profession (see, e.g., [20]).

---

[1] Geoffrey Hinton received the Association for Computing Machinery A.M. Turing Award – the computer science equivalent of a Nobel Prize – in 2018, together with Yoshua Bengio and Yann LeCun, for their contributions to the advancement of DL AI.

DOI: 10.1201/9781003110767-2

More than five years have passed, and Geoffrey Hinton's predictions have not come true: radiologists are still as busy and essential to the healthcare ecosystem as ever. The adoption of AI in radiology has been strengthened, refined, discussed, and expanded, largely thanks to the radiology community. Radiologists quickly and decisively moved from a place of potential fear and alarmism to a position of leadership. Today, there is a consensus within the radiology community that AI will not replace radiologists; instead, it will augment radiologists' abilities and assist them in some of their tasks.

Indeed, the workload for radiologists has dramatically increased, with a staggering 300% surge in imaging volume in recent years. However, the number of trained radiology residents is not proportionately growing to meet this heightened demand. The resulting job market dynamics are exceptionally intense. This situation underscores the urgency and necessity of AI implementation in radiology. Through AI, we can enhance the efficiency of existing radiologists, enabling them to manage the escalating workload effectively and accurately.

Beyond merely utilizing the latest AI tools, the radiology community has further expanded its engagement with AI and DL. This includes involvement in areas such as

- *Developing AI-powered tools*: Radiologists have developed AI tools that can help them diagnose diseases more accurately and efficiently.
- *Training radiologists on AI*: Radiologists are now being trained on AI concepts, techniques, algorithms, and frameworks. This training helps radiologists to understand how AI works and how to use it to improve their diagnostic skills.
- *Collaborating with developers*: Radiologists collaborate with software developers to develop new AI-powered solutions. This collaboration helps to ensure that these solutions are developed in a way that meets the needs of radiologists.

This chapter showcases initiatives by the radiology and imaging informatics communities. They are actively addressing the need for medical professionals to "know enough AI to strip away the mystery" of AI and DL solutions for radiology [21].

2.2 AI in Radiology: Opportunities and Applications

AI is rapidly transforming the field of radiology. AI-powered tools are being developed to help radiologists with various tasks, from image interpretation to diagnosis to patient management.

AI applications in radiology span a vast spectrum. They start when an imaging study is requested, where AI can help radiologists decide which imaging studies to order for their patients. For example, AI can analyze a patient's electronic health record and identify risk factors for certain diseases. This information can then be used to recommend specific imaging studies most likely to help diagnose the patient's condition.

Once the images have been acquired, AI comes into play again to support image analysis.[2]

DL algorithms can identify and highlight potential abnormalities, aiding radiologists in their interpretations. AI algorithms can also quantify specific image features that can be challenging for the human eye to detect, adding an extra layer of precision to the diagnostic process.

AI tools can also facilitate triaging cases, automatically prioritizing those that contain urgent findings and require immediate attention. This ensures that time-sensitive diagnoses are not missed in a busy workflow.

---

[2] See Chapter 4.

Finally, once the analysis is complete, AI can assist in generating preliminary reports, streamlining the communication process to referring physicians. These reports can help synthesize complex imaging findings into concise, clinically relevant information, helping to guide patient management decisions. In this way, AI applications are woven into each step of the radiology workflow, augmenting the abilities of radiologists and improving patient care.

2.3 The AI in Radiology Community Today

In this section, we review some examples of active involvement of the radiology community[3] in advancing AI applications and solutions. We discuss how (and why) some medical doctors took up programming, engaged a community of medical professionals and software developers in search of practical, concrete solutions to meaningful problems, and along the way created numerous mechanisms for the dissemination of scientific knowledge and preparation of the next generations of AI in radiology professionals.

2.3.1 Doctors Who Write Code

Many radiologists took the challenge of writing code and building their solutions to challenging AI problems. They did so for several reasons, among them

- *Satisfying their intellectual curiosity*: Many radiology professionals have an educational background that extends far beyond their MD training and subsequent specializations and certifications in radiology. It is not uncommon to find prominent

---

[3] When we refer to *radiology community*, we implicitly include many other groups of professionals in the medical imaging informatics community.

radiologists with advanced degrees in engineering, computer science, physics, statistics, and other scientific fields.

- *Facing the threat allegedly posed by AI objectively and assessing if it deserves all the hype*: This is brilliantly summarized in a cartoon by Felipe Kitamura, MD, PhD (Figure 2.1), where he indicates that his fear (as a radiologist) of being replaced by AI dropped to zero the day he trained his first AI model. Understanding first hand how brittle and limited DL AI solutions can be provides extremely valuable insight into the hype and ideas on how to leverage this new skill set to build solutions to specific problems.

- *Digging into the inner working of DL AI solutions*: To truly grasp the capabilities and limitations of AI, a deep practical understanding can only be acquired through hands-on experience — writing, running, and modifying code. While a book like this one provides valuable conceptual insights, it is through

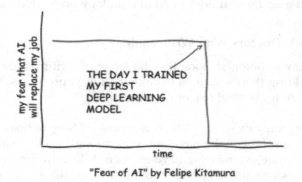

"Fear of AI" by Felipe Kitamura

Figure 2.1
"Fear of AI" (by Dr. Felipe Kitamura, used with permission).

direct engagement with modern languages and frameworks that one can genuinely explore AI's intricacies. Fortunately, with the increasing availability of libraries and toolboxes, including "low-code" and "no-code" tools, training your AI models has become relatively effortless, as these software packages and apps allow you to focus on exploring the core concepts without getting bogged down in the details.[4]

- *Establishing a path for training colleagues and students and inspiring the community to tackle challenging problems together.*

Once radiologists had their intellectual satisfaction satiated, their fear of replacement and obsolescence dispelled, and a greater understanding of the inner workings of AI solutions acquired, they immediately moved on to create a path for engaging colleagues in this new era of their professional lives, as we see next.

2.3.2 Dissemination of Scientific Knowledge

Radiologists have been actively promoting AI efforts and have created and expanded the opportunities for building, sharing, and publishing their knowledge in this area. Following are some representative examples.

2.3.2.1 Journals, Blogs, Podcasts, and Social Media

In January 2019, the Radiological Society of North America (RSNA) launched a new scientific journal, *Radiology: Artificial Intelligence*, focused on "the application of emerging tech in radiology," with coverage of topics such as "the impact of artificial intelligence (AI)

---

[4] See Chapters 8 and 10 for additional discussions and resources.

on diagnostics and patient care, AI's role in radiology education, and the ethical implications that AI presents in the medical field" [22]. The journal's editor also maintains an active blog, "The Vasty Deep" [23], and promotes monthly live one-hour discussions on Twitter (under the #RadAIchat hashtag), which are eventually summarized and archived. A companion podcast [24] for sharing the advances in AI in radiology was launched in April 2020.

The American College of Radiology (ACR) also has a rich blog [25], launched in late 2018, whose posts – written by a broad range of contributors – cover topics of interest on AI in radiology.

2.3.2.2 Challenges and Competitions

During the past few years, prominent societies such as the RSNA, the Medical Image Computing and Computer Assisted Intervention Society (MICCAI), the ACR, the Society of Thoracic Radiology, the Foundation for the Promotion of Health and Biomedical Research of Valencia Region, the International Skin Imaging Collaboration, the American Society of Neuroradiology, the Medical Imaging Databank of the Valencia Region, and the Society for Imaging Informatics in Medicine have created open challenges – often in partnership with the popular ML platform Kaggle [26] – with publicly available datasets and prize money.

These competitions have covered the use of AI/ML/ DL techniques to solve problems such as brain tumor radiogenomic classification [27], intracranial hemorrhage detection [28], pulmonary embolism detection [29], COVID-19 detection from chest x-rays [30], pneumonia detection from chest radiographs [31], melanoma classification [32], pediatric bone age estimation [33], pneumothorax segmentation from chest x-rays [34], detection of fractures in cervical spine computed tomography images

[35], and detection of suspected breast lesions in mammography images [36].

These efforts are in addition to numerous other challenges and competitions from the medical image analysis community, e.g., The Grand Challenge [37], the MICCAI challenges [38], and ImageCLEF [39] medical tasks. We return to this topic in Chapter 8.

2.3.2.3 Hands-On Training and Other Efforts

Some of the most prominent annual meetings and conferences in radiology and medical imaging informatics worldwide have started to include hands-on courses and "learning labs" in AI and DL topics, ranging from beginner classes to highly specialized modules. These training programs are usually led by "doctors who can code."

Another unique example of the blend between AI and radiology is the "Magician's Corner" column [40], a feature of the *RSNA Radiology: AI* journal. The column, led by Dr. Bradley Erickson, RSNA volunteer leader for the Medical Imaging Deep Learning Classroom, focuses on articles that "seek to demystify the tools, processes, and results of deep learning." The companion code for the examples presented in the column is publicly available on GitHub [41].

2.3.3 Preparing the Next Generation

One of the most apparent indicators that the radiology community was embracing change and willing to be in the driver's seat as AI advancements in radiology became more frequent and prominent was recognizing the need to create AI/ML curricula and training programs for radiologists. This sentiment was clearly expressed by Wood et al. in their May 2019 paper, where the authors stated that "rather than succumb to fear and skepticism, future radiologists must be equipped with a working knowledge of ML to leverage the tools as they are deployed" [42]. In Chapter 7, we expand the discussion and provide concrete recommendations for integrating AI in radiology education at the undergraduate and graduate levels.

In summary, AI in radiology has evolved dramatically in the past few years, thanks to a series of bold efforts by the radiology community to dispel fears of radiologists being replaced by AI and along the way develop numerous opportunities for active engagement and training of radiology professionals in topics related to AI, DL, and ML.

Key Takeaways

- The radiology community has swiftly and effectively embraced AI and DL, moving from fear and alarmism to a consensus that AI will augment radiologists' abilities rather than replace them.

- Radiologists, medical imaging informatics professionals, and AI experts have taken an active role in disseminating scientific knowledge related to AI in radiology. They have created and expanded opportunities for building, sharing, and publishing their expertise in this area, contributing to the overall advancement of AI in radiology.

- The recognition of the need for radiologists to have a working knowledge of ML reflects the community's commitment to being in the driver's seat as AI advancements in radiology become more frequent and prominent. This proactive approach ensures that radiologists can leverage AI tools effectively and stay at the forefront of their field.

3

Fundamentals of Machine Learning and Deep Learning

3.1 Introduction

Machine learning (ML) and deep learning (DL) reprsent two critical realms of artificial intelligence (AI) that are driving innovations across various sectors, including healthcare. ML uses algorithms that iteratively learn from data, allowing computers to find hidden insights without being explicitly programmed to look for specific patterns or features. ML algorithms can analyze vast amounts of data and deliver fast, accurate results to facilitate informed decision-making.

DL is a subset of ML that uses artificial neural networks, which are computational models inspired by the human brain and capable of recognizing patterns in unstructured data such as images, sound, and text. In contrast to traditional ML algorithms that improve performance with manual feature extraction and fine-tuning, DL models automatically learn features and improve their performance with the increase in data size and computational resources.

This chapter provides an overview of ML and DL principles and how they apply to radiology.

DOI: 10.1201/9781003110767-3

3.2 Machine Learning

ML is a branch of AI that focuses on the design and implementation of algorithms that acquire knowledge directly from data. Unlike traditional computing approaches, where solutions are based on predetermined equations or explicitly defined rules, ML algorithms self-learn patterns from the data they're exposed to.

The ML process is typically divided into two major phases: the *training* phase and the *inference* phase.

In the training phase, algorithms learn patterns from a dataset known as the *training set*. This dataset comprises examples pertinent to the problem at hand, with each example typically consisting of a set of features and a corresponding label or target variable. In supervised learning, one of the most common forms of machine learning, these labels represent the desired prediction or output. The learning aspect involves the algorithm iteratively making predictions on the training data and adjusting its internal parameters based on the difference between its predictions and the actual labels, also known as the *error*. The goal of this phase is to minimize this error, resulting in a model that can accurately map the relationship between features and labels.

Following training, the model undergoes the inference phase. Here, the learned model is applied to new, unseen data to generate predictions or decisions. This *test set* is distinct from the data used during training. The model's performance during the inference phase offers a valuable indicator of its ability to generalize from learned patterns to new instances, which is vital in practical applications.

Evaluating the performance of ML models is a crucial aspect of developing AI applications, including those in radiology. Several metrics can be employed to quantify the efficacy of these models, each offering a unique perspective on the model's behavior, including

- *Accuracy*, which measures the proportion of predictions a model gets right. Although accuracy is a straightforward metric, it doesn't provide a comprehensive view of the model's performance, particularly when dealing with skewed or imbalanced datasets. In those cases, *sensitivity*, also known as true positive rate (TPR), or *specificity*, also known as true negative rate (TNR), might be preferred.

- The *receiver operating characteristic (ROC) curve* (Figure 3.1), which is a graphical representation of the model's diagnostic ability. It plots the TPR (sensitivity) against the false positive rate (1–specificity) at various threshold settings. The *area under the ROC curve (AUC)* provides a scalar measure of the model's performance across all thresholds, with a value of 1.0 representing a perfect model and 0.5 indicating a model performing no better than random chance.

There are five primary types of ML:

1. In *supervised learning*, the ML algorithm observes some labeled examples consisting of input-output pairs and learns a function that maps from input to output. Supervised learning is predominantly employed in scenarios where ample labeled data are available.

2. In *unsupervised learning*, the ML algorithm learns patterns using unlabeled data as input, which makes the task more about finding structure in the data rather than making accurate predictions. Unsupervised learning algorithms discover hidden patterns or data groupings without the need for human intervention.

3. In *semisupervised learning*, the ML algorithm is given a few labeled examples and uses the learned

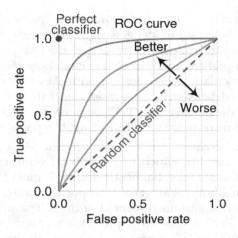

Figure 3.1
Receiver operating characteristic (ROC) curve (from Wikimedia Commons).

knowledge to handle a large collection of unlabeled examples. Semisupervised learning can be used to improve the performance of ML models when there is a limited amount of labeled data available.

4. In *self-supervised learning*, the ML algorithm learns representations from data by predicting certain parts of it without relying on explicit human-provided labels. Instead, it uses the inherent structure of the data to create its own labels.

5. In *reinforcement learning*, the ML algorithm[1] learns to make sequential decisions through trial-and-error interactions with an environment, aiming to maximize cumulative rewards by adjusting its actions based on received feedback signals that can act as rewards or punishments. The algorithm's goal is to choose actions that maximize the cumulative reward over time.

---

[1] In the context of reinforcement learning, an ML algorithm is often referred to as an *agent*.

Most contemporary ML applications use *supervised* learning techniques[2] applied to two main types of tasks:

1. *Classification*: The task of predicting a discrete class label, for example, in the context of radiology, ML algorithms can be trained to distinguish between *normal* and *cancerous* breast tissue in mammograms. Examples of ML classification techniques include naive Bayes, k-nearest neighbors, logistic regression, support vector machine (SVM), decision trees, and neural networks.

2. *Regression*: The task of predicting a continuous quantity, for example, given an image of a radiography of a child's hand and wrist, an ML algorithm can be trained to estimate their age. This can help doctors understand if a child's bone growth is progressing normally or if there might be any growth disorders. Examples of regression techniques include linear regression, decision trees, and neural networks.

Selecting which ML algorithm to use for a specific problem can be overwhelming. There are dozens of supervised and unsupervised ML algorithms in the literature, and each technique takes a different approach to learning. In other words, there is no "best method" or "one size fits all." Finding the right algorithm is often an iterative process where a fair amount of trial and error should be expected. The algorithm selection process also depends on many application-specific factors, such as the size and type of data available, the insights that the resulting model should get from the data, and how those insights will be used within the context of the application.[3]

---

[2] A detailed explanation of these techniques is beyond the scope of this book. See Chapter 10 for useful resources.

[3] The scikit-learn site [43] provides a (highly simplified) flowchart to help beginners navigate the space of candidate ML algorithms for a given problem.

3.3 Deep Learning

During the past few years, most of the news revolving around advancements in AI came associated with the phrase *deep learning*. DL is the umbrella term used to describe techniques that use very large and complex neural networks to solve complex tasks involving different types of data (text, images, audio, video, sensor readings, etc.). In this section, we look at the basics of DL, including some of the most popular DL architectures and approaches.

3.3.1 Artificial Neural Networks: The Building Blocks of Deep Learning

An artificial neural network consists of interconnected layers of nodes, or "neurons," where each layer transforms its input into a more abstract representation. These layers are typically categorized into three types: an *input layer* that receives the data, one or more *hidden layers* where the data are processed, and an *output layer* that delivers the final decision or prediction (Figure 3.2). Each neuron within a layer is connected to neurons in the next layer through *edges* that carry *weights*, representing the influence one neuron has on another. *Activation functions* introduce nonlinearity into the neural network, allowing it to learn from complex and nuanced data. They dictate whether a neuron should be activated based on the weighted sum of its input.

Neural networks can be trained to learn from data by adjusting the weights of the connections between the nodes based on the error between the network's output and the desired output. This error is calculated using a *loss function*, which is a mathematical function that measures the

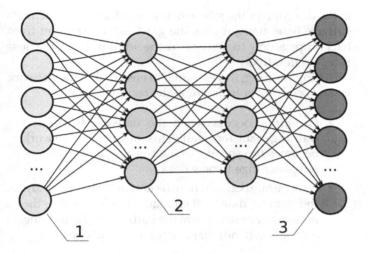

Figure 3.2
Typical artificial neural network architecture: (1) input layer, (2) hidden layers, (3) output layer (from Wikimedia Commons. See https://commons.wikimedia.org/wiki/File:Neural_network_bottleneck_achitecture.svg.)

difference between the network's output and the desired output.

Backpropagation is an algorithm used for calculating the gradient, or the derivative of the loss function, with respect to the weights in the neural network. This process involves propagating the error of the model's predictions backward through the network, from the output layer to the input layer. The calculated gradient represents how much the loss function will change if we change the weights by a small amount. Therefore, backpropagation provides information about the *direction* we need to adjust the weights to reduce the loss function.

Gradient descent is an optimization algorithm that uses the gradients calculated by backpropagation to adjust the weights in the network. It does so by moving the weights in the direction of the steepest descent of the loss function's surface. Essentially, it uses the gradient information from backpropagation to *descend* to the minimum of the loss function iteratively.

During training, it is essential to ensure that the network is not *overfitting* or *underfitting* the training data:

- *Overfitting* occurs when the network learns the training data too well. This means the network becomes too specific to the training data and cannot generalize to new data.

- *Underfitting* occurs when the network does not learn the training data well enough. This means that the network cannot capture the patterns in the training data and will not make accurate predictions.

The *validation set* plays a crucial role in preventing overfitting. After a model is trained on the training set, it is evaluated on the validation set, which contains different data than the training set but follows the same distribution. The model's performance on the validation set indicates how well the model generalizes to unseen data. If a model performs well on the training set but poorly on the validation set, it's a sign that the model might be overfitting to the training data. This feedback can be used to adjust the model's complexity or tweak its hyperparameters.[4] Once the model's hyperparameters are adjusted, and its performance is satisfactory on the validation set, it is finally evaluated on the test set. This gives a final measure of how the model is expected to perform on completely unseen data in the real world.

---

[4] In the context of deep neural networks, *hyperparameters* are parameters that control the architecture of the network and the learning process. They are set before the training process begins. Some examples of hyperparameters include the learning rate (which determines how quickly the network updates its weights during training) and the batch size (which determines how many training examples are processed at a time).

3.3.2 Deep Learning Versus Traditional Machine Learning

DL and ML are both branches of artificial intelligence that allow algorithms to learn from data and make predictions. Both techniques can be used to solve various problems, including classification and regression. However, they differ in their approach to learning. Traditional ML algorithms typically use handcrafted features to learn from data, while DL algorithms use neural networks to learn features automatically.

Some of the key differences between DL and traditional ML include

- *Level of human intervention*: In traditional ML, features are typically engineered manually, which requires domain expertise and can be timeconsuming. DL, however, learns to extract useful features directly from the data, reducing the need for manual feature extraction.

- *Choice of technique*: Traditional ML algorithms can use various techniques, including decision trees, support vector machines, and naive Bayes classifiers. DL algorithms use artificial neural networks, which vary significantly in size, shape, architecture, and complexity [44].

- *Data dependency*: DL algorithms require more data to learn effectively than traditional ML models.

- *Computational resources*: DL models are typically more computationally intensive and require more processing power than traditional ML models due to the complexity and size of the neural networks.

- *Quality of results*: ML techniques typically achieve good results for tasks with less complexity, while DL is better suited for tasks with greater complexity.

- *Interpretability*: Traditional ML models are often more interpretable than DL models. The latter are frequently referred to as "black boxes" due to their complex internal mechanisms.

3.4 Selected Deep-Learning Architectures

In this section, we provide a brief overview of the most widely used DL architectures in medical image analysis and radiology. The main goal of this section is to provide a top-level understanding[5] of representative architectures and the types of problems they can solve.

3.4.1 Convolutional Neural Networks

Convolutional neural networks (CNNs) are among the most widely used DL architectures for medical image analysis. CNNs are well-suited for image classification tasks because they can learn the spatial relationships between pixels in an image. This makes them particularly useful for image classification, object detection, and segmentation [45].

This architecture has multiple layers, including convolutional layers, pooling layers, and fully connected layers, which work together to extract features from input images and classify them. The convolutional layers apply a series of filters to the input data, allowing the model to automatically learn features from raw pixel data. Pooling layers then reduce the spatial size of the data, minimizing

---

[5] A deep understanding of these architectures is significantly beyond the scope of this book and may be a long and challenging undertaking. See Chapter 10 for suggestions of learning resources.

computational demand and controlling overfitting. Finally, the fully connected layers aggregate the high-level features learned by previous layers to classify the image (see Figure 3.3).

CNNs are employed in various applications within radiology, including the detection and diagnosis of diseases like cancer, the classification of lung nodules, and the segmentation of organs in computed tomography (CT) scans.

3.4.2 Recurrent Neural Networks

A recurrent neural network (RNN) is a type of neural network architecture designed to process sequential data, allowing information to persist over time through recurrent connections.

While CNNs are primarily used for image processing tasks, RNNs are better suited for sequential data such as text, speech, or time series. Unlike CNNs, RNNs have feedback connections that enable them to capture dependencies and patterns over arbitrary time steps, making them more suitable for tasks with temporal dynamics.

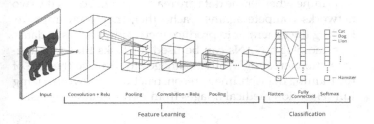

Figure 3.3
A typical CNN allows end-to-end training, taking raw image pixels as input and producing a result (in this case, the predicted label in an image classification task) at the output.

In radiology, RNNs can be helpful in analyzing radiology reports, where they can extract and classify information. They are also employed in multiframe image sequences like cardiac magnetic resonance imaging (MRI), where the temporal sequence matters.

3.4.3 Long Short-Term Memory Networks

Long short-term memory networks (LSTMs) [46] are a type of RNN specifically designed to avoid the long-term dependency problem, where the network fails to learn from important events that happened many steps back in the sequence. They can remember patterns over relatively long sequences, making them useful for tasks involving longer sequences.

Similarly to RNNs, LSTMs have shown great promise in tasks that involve sequential data. For instance, they could be used to analyze a series of radiology reports and predict the progression of a disease over time.

3.4.4 Generative Adversarial Networks

Generative adversarial networks (GANs) [47] consist of two neural networks: a *generator* and a *discriminator*. The generator produces synthetic data, and the discriminator tries to determine whether the data are real or generated. As the two networks compete against each other, they both improve, leading the generator to produce increasingly realistic data.

In radiology, GANs can be used for tasks like data augmentation (creating synthetic images to increase the size of the training dataset), image reconstruction, and synthesizing full-resolution medical images from lower-quality inputs.

3.4.5 Transformers

Transformers [48] are a type of neural network architecture that is used for natural language processing (NLP) tasks. Transformers are based on the attention mechanism, which allows them to learn long-range dependencies between words in a sequence. This makes them particularly well-suited for tasks such as machine translation, text summarization, and question answering.

Transformers are composed of two main components: an encoder and a decoder. The encoder takes a sequence of words as input and produces a sequence of hidden states. The decoder then takes these hidden states as input and produces a sequence of output words.

The attention mechanism is used in both the encoder and the decoder. In the encoder, the attention mechanism allows the transformer to learn which words in the input sequence are most relevant to each other. In the decoder, the attention mechanism allows the transformer to learn which words in the output sequence are most relevant to the words in the input sequence.

BERT, which stands for Bidirectional Encoder Representations from Transformers [49], is a groundbreaking model in the field of NLP. BERT uses only the encoder component of the transformer architecture. It is designed to pre-train deep bidirectional representations from unlabeled text by encoding the context of "tokens" (e.g., words) on either side of a target token in a stream of text (e.g., a sentence in a document). After pre-training, BERT can be fine-tuned on a specific task – such as text classification, question answering, or named entity recognition – and domain-specific corpora, which may result in improved performance on tasks such as analysis of radiology reports [50].

Language-based transformer architectures are becoming increasingly popular in radiology, as they have been shown to be effective for a variety of tasks, including

- *Text classification*: Transformer models, such as BioBERT [51], have been shown to be effective for classifying different types of medical reports, such as radiology reports and discharge summaries.

- *Answer extraction*: Transformer models have been used to extract answers to clinical questions from radiology reports [52].

- *Summarization*: Transformer models have also been used to summarize radiology reports in a concise and informative way [53].

Large language models (LLMs) such as Bard and generative pre-trained transformers (GPT) are transformer-based models trained on massive datasets of text and code. This allows them to learn the statistical relationships between words and phrases and to generate text that is both coherent and grammatically correct. Such LLMs are referred to as *foundation models* because they can be used as the foundation to train models for more specific downstream tasks. Radiology-specific LLMs, such as Radiology-GPT [54], have recently been introduced, and the field is expected to grow significantly in the next few years.

3.4.6 Vision Transformers

Vision transformers (ViTs) [55] are a type of transformer-based model used for computer vision tasks. They are based on the attention mechanism, which allows them to learn long-range dependencies between pixels in an image. This makes them particularly well-suited for tasks such as image classification, object detection, and segmentation.

ViTs are being used in a variety of medical imaging tasks, including

- Image classification using different types of medical images, such as x-rays, CT scans, and MRIs
- Detection of tumors, lesions, and other abnormalities in medical images
- Segmentation of organs and tissues in medical images

3.5 Transfer Learning

In the context of AI for radiology, *transfer learning* [56] is an extremely valuable technique that leverages pre-trained DL models to address new tasks. It can be used for image- as well as text-based tasks.

3.5.1 Transfer Learning for Imaging Tasks

Given that medical image datasets can be limited in size due to privacy considerations and the difficulty of obtaining labeled data, using models that have been pre-trained on large, diverse datasets (often non-medical, such as ImageNet [57], a dataset containing 1 million general-purpose images from 1,000 different categories) can provide a strong initial knowledge base, helping overcome the limitation of small dataset sizes.

The principle behind transfer learning is that these pre-trained models have already learned to identify various features from the dataset they were trained on, and these features can serve as a helpful starting point for new tasks. For instance, a model trained on a large dataset of everyday images will have learned to recognize shapes, textures, and patterns that can also be relevant in medical imaging.

In practice, transfer learning often involves taking a pre-trained model and replacing and retraining the final layer(s) to adapt to the new task (e.g., classifying a specific type of lesion in radiology images). This approach can significantly reduce the computation time and data requirements compared to training a DL model from scratch. Additionally, it often results in improved performance, particularly when the available medical imaging data are limited.

3.5.2 Transfer Learning for Natural Language Processing Tasks

Transfer learning can also be used in the context of NLP, starting from a pre-trained language model, such as BERT or GPT, and fine-tuning it on radiology-specific tasks. These pre-trained language models are generally trained on a large corpus of text data from the internet, allowing them to learn a rich representation of the structure and semantics of the language. They can learn to generate coherent text, answer questions, translate between languages, and much more. When these models are fine-tuned on a specific task, they leverage this general language understanding to perform the task effectively.

For example, consider a radiology-specific task of automatically generating radiology reports given a set of imaging data. A language model can be fine-tuned for this task by training it on a dataset of radiology reports, where it learns to generate text that mimics the style and content of these reports. The model could start with the general language understanding learned from the internet text and then adapt to the specifics of radiology language, including technical terminology, common phrases, and even the structure of a radiology report. This is significantly more efficient than training a model from scratch on the radiology data, as the model can leverage the general language knowledge it has already learned.

3.6 The Machine-Learning/Deep-Learning Workflow

ML and DL solutions are usually structured as a workflow that consists of the following main steps[6]:

---

[6] These steps are essentially a blend between the steps suggested by Chollet [58] and those suggested by Geron [59].

1. *Look at the big picture*: The first step in any ML/DL project involves understanding the problem, the available data, resources, and the project's goals and potential impact.

2. *Define the problem and assemble a dataset*: This involves specifying the project's desired outcome and the criteria that will be used to measure its success. At this step, a large and representative dataset must be built.

3. *Discover and visualize the data to gain insights*: Visualization of the distribution of the data and the relationships between different features is an essential step for understanding and trusting the data and identifying any potential problems with the data.

4. *Choose a measure of success*: This metric will be used to evaluate the performance of the ML/DL model. The chosen metric should be relevant to the problem, and it should be quantifiable.

5. *Decide on an evaluation protocol*: This is a procedure used to evaluate the performance of your model. The evaluation protocol should be repeatable and reliable and designed to test the model's performance on unseen data.

6. *Prepare the data*: This involves cleaning the data, removing outliers, and transforming the data into a format that is compatible with the selected ML algorithm or DL architecture.

7. *Develop a model that does better than a baseline*: This simple model can be used as a starting point for experimentation with different models, their variants, and hyperparameters.

8. *Scale up: develop a model that overfits*: This involves increasing the size of the model and the amount of data that the model is trained on. "The ideal model

is one that stands right at the border between underfitting and overfitting, between under-capacity and overcapacity. To figure out where this border lies, first you must cross it" [58].

9. *Fine-tune your model*: This consists of adjusting the model's hyperparameters to find a set of hyper-parameters that achieves the best performance on the evaluation dataset.

10. *Present your solution*: Communicating the results of the project to stakeholders is an essential mile-stone. The presentation should include an over-view of the project, the results of the project, and the implications of the project.

11. *Launch, monitor, and maintain your system*: This encom-passes deploying the model in production, moni-toring its performance, and updating it as new data become available.

In Chapter 8, we revisit and expand this workflow and provide practical advice on how to get started building your own AI solutions.

In summary, ML techniques and artificial neural net-works are at the heart of the latest impressive devel-opments in AI. ML and DL can be applied to various problems in radiology, including medical image analysis. This is a vast problem space, as discussed in Chapter 4.

Key Takeaways

- ML is a branch of AI that deals with creating algo-rithms that learn directly from data.

- DL is the umbrella term for techniques that use very large and complex neural networks to solve complex tasks involving different data types.

- Artificial neural networks are mathematical constructs that can learn complex patterns directly from the data.

- ML and DL solutions must be trained (i.e., have their internal "knobs" adjusted to reflect that they have, indeed, "learned" what is expected from them) before they can be deployed and used.

- There are tens of different deep neural network architectures in use today, including CNNs, RNNs, LSTMs, GANs, and transformers.

- The choice of architecture depends on the type of problem and the nature, amount, and quality of the data.

- A DL design paradigm known as *transfer learning* allows the design of DL solutions that can leverage the knowledge encoded into pre-trained DL models.

- The ML/DL workflow follows a predictable pattern that starts with carefully investigating the question at hand and the data available to answer it. It ends at the point where the model is deployed and maintained.

4

Fundamentals of Medical Image Analysis

4.1 Introduction

Medical image analysis is an interdisciplinary field that intersects computer science, mathematics, physics, and medicine. It involves processing and interpreting medical images to aid in diagnosis and facilitate therapeutic interventions. By analyzing medical images, healthcare professionals can understand the anatomy and function of the human body in a noninvasive manner, allowing them to detect, diagnose, and monitor diseases more accurately.

At its core, medical image analysis involves using computational methods to extract vital diagnostic information from medical images. This consists of various procedures, including image acquisition, preprocessing, segmentation, feature extraction, and classification. The primary goal of medical image analysis is to improve the quality and interpretability of medical images, thereby aiding healthcare professionals in their clinical decision-making processes.

In this chapter, we provide a brief and broad overview of the expansive field of medical image analysis, covering the primary imaging modalities, providing an overview of the main steps in the image processing and analysis pipeline, and concluding with practical considerations such as interoperability, standardization, and the potential for innovation using artificial intelligence (AI) in this ever-growing research field.

DOI: 10.1201/9781003110767-4

4.2 Medical Imaging Modalities Used in Radiology

Radiological sciences have experienced significant progress during the past decades, thanks to advancements in medical imaging and computer-based medical image processing. Some of the most popular imaging modalities in use today have been developed in the past 50 years. This section provides a brief overview of the most popular imaging modalities used in radiology. Each modality has advantages and disadvantages and specific applications based on the organ or tissue type and the disease condition being investigated.

There are five main types of medical imaging techniques: radiography (commonly referred to as x-ray), computed tomography (CT), magnetic resonance imaging (MRI), ultrasound (US), and positron emission tomography (PET). The main aspects of each modality are described next.[1]

4.2.1 X-ray

X-ray imaging is one of the oldest and most commonly used forms of medical imaging. It involves the emission of a small amount of radiation that passes through the body to capture images of the internal structures. X-ray imaging is fast and relatively inexpensive, making it the first-line imaging method for many conditions, including bone fractures and lung diseases.

X-rays were first discovered by Wilhelm Conrad Röntgen, a German physicist, in 1895. Röntgen's pioneering work revolutionized the field of medicine by providing a noninvasive

---

[1] The technical details behind the physics and electronics of medical imaging equipment are beyond the scope of this text. See Chapter 10 for additional resources.

method to peer inside the human body. For his ground-breaking discovery, Röntgen was awarded the first Nobel Prize in Physics in 1901.

Essentially, x-ray radiography consists of the transmission of x-rays (the name given to electromagnetic radiation in the 0.001 to 0.1 nm wavelength range or frequencies within the 1018 to 1020 Hz range) through the body and collecting the x-rays using a film or an array of detectors. The attenuation or absorption of x-rays depends on the type of tissue that they go through, e.g., bone tissue (high attenuation), soft tissues (medium), or air (low). The attenuation of radiation intensity is determined at each location by measuring the difference in intensity between the source and the detector. A two-dimensional (2D) attenuation map can be recorded on a radiographic film for an x-ray (film) radiograph. Contemporary digital x-ray radiograph equipment performs a similar operation but stores the attenuation digitally using x-ray detectors and electronic instrumentation [60]. Technically, the recorded 2D image is a projection of the three-dimensional (3D) anatomical structure being imaged.

X-ray mammography is a specialized radiographic imaging method used for breast imaging. Due to the soft and vascular nature of breast tissue, the image acquisition process requires compression of the breast to acquire images of enough quality to detect small lesions and microcalcifications.

From the viewpoint of medical image analysis, x-ray radiographs and mammograms are essentially 2D grayscale intensity images whose spatial resolution (number of pixels per unit area) and number of gray levels (number of different shades of gray that can be represented in the image) can vary depending on the type of imaging device used, the settings of the imaging device, and the patient's anatomy.[2]

---

[2] The spatial resolution and number of gray levels of an image are important considerations for medical image analysis. For example, a machine learning algorithm that is trained to classify tumors in mammograms will need to be trained on a dataset of mammograms that have similar spatial resolution and number of gray levels as the mammograms that the algorithm will be used to classify.

AI techniques can be used with x-ray imaging to refine imaging procedures, enhance image quality, and increase diagnostic precision. Recent developments include the application of AI for expediting image acquisition and reconstruction, minimizing noise artifacts, automating image interpretation, and predicting the course of disease progression. Notably, AI has been utilized in chest x-rays for the automated detection of conditions like pneumonia and tuberculosis, significantly reducing diagnostic time and improving patient outcomes.

4.2.2 Computed Tomography

Computed tomography, also known as CT scans, is an advanced imaging technique that uses x-rays in conjunction with computer technology to produce more detailed cross-sectional images of the body. Unlike traditional x-rays, which provide a 2D view, CT scans allow for a 3D view of the body's structures. CT scans are particularly useful for detecting various types of cancers, visualizing complex fractures, and assessing other internal injuries and damage. CT scans have been used in multiple subdomains of radiology, including head and neck, thorax, urogenital tract, abdomen, and musculoskeletal system.

The invention of CT is attributed to Sir Godfrey Hounsfield, an English engineer, and Dr. Allan Cormack, a South African physicist. Hounsfield developed the first commercially viable CT scanner while working at EMI Laboratories in England in the early 1970s. Concurrently, but independently, Cormack provided the mathematical underpinnings to reconstruct images from the raw data produced by the scanner. For their pioneering contributions to the development of CT, Hounsfield and Cormack were jointly awarded the Nobel Prize in Physiology or Medicine in 1979.

The basic principle of x-ray CT is the same as that of x-ray digital radiography: x-rays are transmitted through the body and collected by an array of detectors to measure the total attenuation along the beam path [60]. But instead of producing a 2D projection of the anatomical structure, CT equipment produces a stack of 2D slices which can be combined into a volume corresponding to the 3D structure being imaged through a mathematical process known as *filtered backprojection* or *iterative reconstruction*.

These reconstruction algorithms consider the x-ray source's path and the attenuation of x-rays as they pass through different tissues. The result is a set of images where the pixel values correspond to the linear attenuation coefficient, a property related to the tissue's density and atomic number.

The scanner geometry of CT equipment consists essentially of rotating x-ray sources and associated detectors. The exact arrangement of sources and detectors – and how they rotate – has evolved significantly over time. The literature refers to significant advancements in CT technology as *generations*. Most of today's CT scanners are third generation, which consist of helical (spiral) scanners with a continuous rotating x-ray source and multiple detectors. However, there is a growing trend toward fourth-generation electron beam CT scanners, which offer improved image quality and faster scan times.

In addition to general-purpose CT scanners, several dedicated scanners are in use for tasks such as breast cancer screening, oral and maxillofacial procedures, and many interventional radiology tasks, such as angiography, fluoroscopy, and spine/orthopedic surgery.

CT imaging can be done with or without using contrast agents, such as barium sulfate, administered orally or via intravenous injection. The goal of using contrast agents is to increase the attenuation of areas of interest, leading to improved contrast.

From the viewpoint of medical image analysis, CT scans are essentially 3D grayscale intensity images that can be decomposed into 2D slices (or planes). Each element of the 3D volume is called a *voxel* (volume pixel). The in-plane spatial resolution (number of pixels per unit area in a single slice) and the number of gray levels (different shades of gray that can be represented in a single slice) might vary among different samples, depending on the scanner used and the scanner's settings. The voxels in a CT scan are typically cubic in shape, but they can also be rectangular or cylindrical. The size of a voxel in a CT scan can vary depending on the scanner used and the settings of the scanner. The smaller the voxel size, the higher is the spatial resolution of the CT scan. The number of gray levels in a CT scan can vary from 8 to 12 bits per pixel. The higher the number of gray levels, the more shades of gray can be represented in the CT scan. This can improve the quality of the image and the ability to extract diagnostic information from the image.

The fusion of AI with CT can dramatically optimize imaging processes, enhance the quality of images, and elevate the accuracy of diagnoses. Modern advancements include using AI for swift image acquisition and reconstruction, reduction of image noise, automation of image interpretation, and prediction of the disease trajectory. AI has been especially valuable in CT imaging for stroke patients, where rapid and accurate interpretation can significantly impact the course of treatment and prognosis.

4.2.3 Magnetic Resonance Imaging

MRI is a noninvasive imaging technology that uses a strong magnetic field and radio waves to generate detailed images of the body's internal structures, particularly soft tissues. Unlike x-rays and CT scans, MRI does not involve ionizing radiation. MRIs are particularly useful for imaging muscles, ligaments, and other musculoskeletal system structures, as well as the brain, spinal cord, and other soft tissues. They provide superior soft tissue contrast compared to other imaging modalities.

Similarly to CT, MRI is a tomographic image acquisition method that produces 3D images of the human body. However, instead of resorting to the use of external radiation (such as CT and x-ray radiography), MRI uses the nuclear magnetic resonance (NMR) property of selected nuclei of the matter of the object. The physical phenomenon of NMR was initially discovered and explained independently by Felix Bloch and Edward Purcell in the 1940s. It was only in the early 1970s, however, that Paul C. Lauterbur realized its application to the field of medical imaging, followed by work by Peter Mansfield on the mathematical theory for fast scanning and image reconstruction needed in clinical practice. Lauterbur and Mansfield were jointly awarded the 2003 Nobel Prize in Medicine or Physiology for their discoveries [60,61].

Since the late 1990s, MRI techniques have become widely used in multidimensional imaging of the human body, providing both structural and physiological information about internal organs and tissues. MRI delivers high-resolution images with excellent soft tissue contrast, superior to x-ray CT because of the underlying physics [60].

MRI acquisition is quite different from x-ray and CT imaging. In MRI, a strong magnetic field aligns the protons in the water molecules in the body. Short bursts of radiofrequency (RF) energy then disturb this alignment. When the RF energy is switched off, the protons realign with the magnetic field, emitting signals in the process. These signals are captured by the MRI scanner and used to create an image.

MRI image reconstruction is a sophisticated process involving several stages. The MRI machine captures signals emitted by the body's protons as they realign with a magnetic field following an RF pulse. This raw data, known as k-space data, represents spatial frequency information rather than image space data. The reconstruction process involves applying a 2D Fourier transform to the k-space data. This mathematical algorithm converts the frequency data into spatial data, thereby converting the complex signal data into a cross-sectional image. This image can then be further processed to enhance contrast, suppress unwanted signals, or generate specific views of the anatomy.

MR imaging can be utilized with or without the administration of contrast agents, depending on the specifics of the clinical situation. Contrast agents are special substances that, when introduced into the body, alter the contrast in MR images, highlighting specific areas or tissues. This enhancement can provide additional detail and clarity, aiding in the diagnosis of certain conditions or the characterization of specific tissues or structures.

Recent advances in MRI include

- *Ultrahigh field MRI*: MRI scanners that use a stronger magnetic field compared to conventional systems and can produce images with superior resolution and contrast, allowing for better visualization of small structures and subtle disease processes.
- *Functional MRI (fMRI)*: fMRI is a noninvasive technique that allows for the mapping of brain activity by detecting associated changes in blood flow. Recent advances have improved fMRI's spatial and temporal resolution, making it an increasingly important tool in neuroscience research and clinical neurology.
- *Diffusion tensor imaging (DTI)*: DTI is a type of MRI that enables the measurement of the restricted diffusion of water in tissue to produce neural tract images. Recent advancements in DTI provide

unprecedented views of the brain's white matter, improving our understanding of brain connectivity and disorders affecting the white matter.

- *MRI-guided interventions*: The use of MRI for guiding surgical and minimally invasive interventions is an exciting area of advancement. Real-time MRI guidance during procedures can improve the precision of interventions, enhance safety, and potentially improve patient outcomes.

From the viewpoint of medical image analysis, MRI scans are structurally similar to CT scans, except that the number of gray levels typically can range from 8 to 16 bits per pixel.

The integration of AI techniques with MRI can significantly streamline imaging workflows, improve image quality, and enhance diagnostic accuracy. Recent advances include using AI for faster image acquisition and reconstruction, noise reduction, automated image interpretation, and prediction of disease progression.

4.2.4 Ultrasound Imaging

US imaging, or sonography, uses high-frequency sound waves to produce images of the body's internal structures. During a US exam, a small probe called a transducer is placed on the skin, sending sound waves into the body and receiving echoes as the sound waves bounce off internal structures. These echoes are then converted into images by a computer. US is often used to monitor pregnancies and diagnose conditions affecting the heart, blood vessels, liver, kidneys, and other organs.

US has its roots in the 19th century when Italian biologist Lazzaro Spallanzani first studied echolocation in bats. However, it wasn't until the 1940s that the first practical application of US for diagnostic imaging in medicine was developed. The credit goes to Dr. Karl Theo Dussik, an Austrian neurologist who attempted to locate brain tumors using transmission US in 1942. Further advancements in the 1950s and 1960s, including the development of echo mode and Doppler US by researchers such as Ian Donald and Paul-Jean Couinaud, made US a practical and widely used imaging modality. By the 1970s, real-time US imaging had been developed, and the technology has continued to evolve and improve.

In US imaging, a transducer emits high-frequency sound waves into the body. These sound waves reflect off internal structures and return to the transducer, which then converts these reflected sound waves into electrical signals. The time it takes for the emitted sound waves to reflect back to the transducer determines the distance of the internal structure from the transducer, allowing the US machine to reconstruct an image of those structures. By combining the data from many different pulses sent in various directions, the US machine constructs a 2D image of those structures.

From the viewpoint of medical image analysis, US images are structurally similar to CT and MRI scans, i.e., 2D or 3D grayscale intensity images.

The integration of AI techniques with US imaging has the potential to revolutionize the way US imaging is used in clinical practice. By improving image quality, automating image interpretation, and predicting disease progression, AI can help to improve patient care and save lives. Examples of how AI is being used in US imaging include the use of AI to automate the image acquisition process, improve image quality, interpret US images, identify potential abnormalities, as well as predict the progression of disease based on US images.

4.2.5 Positron Emission Tomography

PET scans are a type of nuclear medicine imaging that uses small amounts of radioactive material to diagnose or monitor various diseases, including many types of cancers, heart disease, and neurological disorders. A PET scanner detects the gamma rays produced by the radioactive material and creates images that show where the material has gathered in the body.

PET was first developed in the 1960s. The concept of PET scanning was introduced by James Robertson and his associates at Brookhaven National Laboratory in 1961. Their invention, a rudimentary PET device, was used to study brain metabolism and blood flow. However, it was Michel Ter-Pogossian, Michael E. Phelps, Edward J. Hoffman, and others at the Washington University School of Medicine who are credited with developing the first fully ringed PET scanner in the mid-1970s, which led to the modern design of PET machines. Notably, the development of PET scanning was deeply tied to the advancement of radiopharmaceuticals, particularly the use of fluorodeoxyglucose (FDG). This radiolabeled glucose molecule has become the standard radiotracer used in PET scans. Over the years, PET technology has continually improved, with advancements in detector technology, image reconstruction techniques, and the integration of PET with other imaging modalities such as CT and MRI.

In PET imaging, a small amount of radioactive tracer is introduced into the body. This tracer accumulates in areas of high chemical activity, often corresponding to disease areas in the body. The PET scanner detects the gamma rays emitted by the tracer. It uses this information to generate a 3D map of the tracer concentration in the body, which is then reconstructed into an image.

PET detectors capture these gamma rays, and from the nearly simultaneous arrival of the two gamma rays at different detectors, the scanner can determine the original positron's location along a line (line of response). The scanner then uses a reconstruction algorithm (e.g., filtered back projection or iterative reconstruction) to convert the lines of response from many annihilations into a 3D image of the tracer distribution in the body.

The integration of AI techniques with PET imaging has the potential to significantly streamline imaging workflows, improve image quality, and enhance diagnostic accuracy. Recent advances include using AI for faster image acquisition and reconstruction, noise reduction, automated image interpretation, and prediction of disease progression.

4.2.6 Review and Comparison of Different Imaging Modalities

Each imaging technique offers unique advantages and potential drawbacks, and the choice of one over another depends on the clinical question, patient considerations, and available resources. But, most importantly, in the context of this book, deep-learning AI techniques are effective in improving the accuracy of diagnosis for all five of these medical imaging modalities.

X-ray imaging, being the most accessible and cost-effective, is commonly used for initial assessments, especially for skeletal and chest imaging. However, it is less effective for imaging soft tissues. Deep-learning AI has been used to improve the accuracy of x-ray diagnosis, particularly for diseases such as pneumonia and cancer.

CT scans, although involving a higher dose of radiation than x-rays, provide a more detailed view of the body, making them useful for complex cases that require high-resolution images of internal organs or bones. However, their use may be limited in patients who cannot tolerate the required contrast material or who need to limit their radiation exposure. Deep-learning AI has been used to improve the accuracy of CT diagnosis, particularly for detecting small tumors.

MRI offers the best soft tissue contrast among all the modalities and is excellent for neurological, musculo-skeletal, and cardiovascular imaging. However, MRIs are expensive, time-consuming, and unsuitable for patients with implanted medical devices like pacemakers. Deep-learning AI has been used to improve the accuracy of MRI diagnosis, particularly for detecting brain tumors and multiple sclerosis.

Ultrasound is safe, radiation-free, and can provide real-time dynamic images, making it ideal for obstetric and some cardiovascular applications. However, its image quality can be significantly affected by the patient's body habitus and the presence of air or bone in the path of the sound waves. Deep-learning AI has been used to improve the accuracy of ultrasound diagnosis, particularly for detecting fetal anomalies and heart defects.

PET scans provide functional information and are mainly used for cancer diagnosis and monitoring. However, they expose the patient to radiation and are unsuitable for detailed anatomical assessments. Deep-learning AI has been used to improve the accuracy of PET diagnosis, particularly for detecting early-stage cancer.

4.3 Medical Image Analysis: Overview

Medical image analysis can be seen as a specialized sub-field within image processing and computer vision (IPCV), two disciplines that have been around for more than 50 years. Most IPCV techniques can be adapted, modified, or extended to medical imaging tasks.

There is no universal terminology to delimit the boundaries between *image processing, image analysis,* and *computer vision.* We adopt the following convention:

- *Image processing*: operations where the input is an image[3], and the output is a modified version of the image. Examples of techniques and algorithms in this category include denoising, sharpening, deblurring, and pseudocoloring.

- *Image analysis*: operations where the input is an image, and the output is a *labeled image*, where specific regions, edges, or contours from the input image have been outlined. Examples of techniques and algorithms in this category include semantic segmentation, corner detection, and edge extraction, among many others.

- *Computer vision*: operations where the input is an image that is used by an algorithm to perform tasks, such as object detection, object recognition, image classification, object tracking across multiple frames of a video sequence, as well as answer questions related to the semantic contents of the image/video, e.g., how many people appear in a photo or where it was taken.

The impact of deep-learning techniques on IPCV since the successful image classification results obtained by Krizhevsky et al. in ImageNet 2012 [62] has been so significant that there are deep learning–based[4] versions of virtually every IPCV task, often achieving state-of-the-art

---

[3] Depending on the algorithm and problem domain, the input could also be a series of (2D or 3D) images and/or one or more videos. This is valid for all three categories described here.

[4] Since deep learning is a subset of machine learning, the field of computer vision using machine-learning techniques is often referred to as CVML.

results for that task.[5] The IPCV pipeline usually consists of several operations that are chained together in a meaningful sequence. For example, an image may be filtered and cropped before being used as input to an object classifier. The next section describes such a pipeline in the context of medical image processing and analysis.

4.4 The Medical Image Processing and Analysis Pipeline

A typical medical image processing and analysis pipeline consists of several stages. Each of these stages has its purpose and contributes to the overall goal of extracting meaningful information from the images to aid in diagnosis or treatment. These are the main stages:

1. *Image acquisition*: The first step involves acquiring the medical image using modalities such as x-ray, MRI, CT, US, or PET (Section 4.2).

2. *Preprocessing*: In this stage, images are processed to remove noise, enhance contrast, or standardize the image, making it ready for further analysis.

3. *Image segmentation*: This step involves partitioning the image into different regions, usually to distinguish the areas of interest from the background.

4. *Feature extraction*: Once the image has been segmented, it's often necessary to quantify certain characteristics, or "features," of the regions of interest. In the context of deep learning, feature

---

[5] It is commonly said that the history of computer vision will be written in two volumes: (i) before deep learning (mid-1950s to 2012) and (ii) after deep learning (2012 to present).

extraction is often implicitly performed by the model itself.

5. *Prediction*: After relevant features have been extracted, they can be used to make predictions. This might involve determining whether a tumor is malignant or benign, predicting the progression of a disease, or classifying the type of tissue present in the image. In deep learning, this step is often integrated with feature extraction.

6. *Post processing*: This step involves refining and validating the results from the prediction stage. It may involve enhancing the visual aspects of the results or applying a threshold to prediction probabilities.

7. *Visualization and interpretation*: The final stage consists of presenting the results in a way that a physician or radiologist can easily interpret. This could involve overlaying the results on the original image, providing confidence scores, or visualizing the progression of a disease over time.

It is important to note that these stages aren't strictly sequential, and often there is a need to iterate and refine based on the results at each stage. Moreover, the exact steps and methods used can significantly vary depending on the specific problem, the imaging modality, and the available data. Steps 2 through 7 are described in further detail next.

4.4.1 Preprocessing of Medical Images

Medical imaging data require certain preparatory steps before applying AI algorithms or other analysis techniques. These are some of the most common preprocessing steps used in the field of medical image analysis:

- *Image normalization*: Image normalization is a vital step that aims to standardize the pixel intensities across different images, thereby reducing the variability in the dataset. This is particularly important when dealing with medical images as they can vary significantly in terms of contrast, brightness, and dynamic range due to differences in scanning parameters, devices used, and individual patient characteristics. Image normalization can involve several techniques, such as histogram equalization, z-score normalization, or scaling the pixel intensities to a specific range.

- *Noise reduction*: Noise reduction techniques are employed to eliminate or reduce noise artifacts present in the images, which might have been introduced during image acquisition. These artifacts can degrade the image quality and affect the accuracy of subsequent image analysis. Noise reduction can be achieved using various methods, such as smoothing filters, median filters, and more advanced techniques like anisotropic diffusion and nonlocal means filtering. The aim is to reduce noise while preserving the essential features and details in the image.

- *Image enhancement*: Image enhancement techniques are utilized to improve the visual clarity and perceptibility of features of interest in a medical image. This step is crucial because it directly impacts the performance of subsequent stages, such as segmentation and feature extraction. Techniques used for image enhancement may include contrast stretching, edge enhancement, and histogram equalization.

- *Image resizing*: Medical images often come in different sizes, resolutions, and aspect ratios, especially when they are sourced from different types of scanning equipment or protocols. To create uniformity across the dataset, images are often resized to a standard dimension during preprocessing.

- *Image augmentation*: Image augmentation is a strategy used to artificially expand the size of the training dataset and increase its diversity, thereby reducing overfitting and improving the generalization ability of AI models. Common image augmentation techniques include rotation, scaling, translation, flipping, and the addition of noise.[6] More recently, advanced methods such as generative adversarial networks (GANs) are being used to create synthetic medical images for augmentation.

Each of these preprocessing steps plays a different role in enhancing the reliability and performance of medical image analysis solutions. Thus, it is imperative to understand their impact on the images and tailor these steps based on the specific demands of each task.

4.4.2 Segmentation of Medical Images

Image segmentation is a crucial process in IPCV that divides a digital image into multiple segments, or sets of pixels, by assigning a label to every pixel in an image such that pixels with the same label share certain visual characteristics. In the context of medical imaging, these characteristics could represent a particular tissue type, pathology, or anatomical structure. Some of the most popular image segmentation techniques include

- *Thresholding*: This technique selects an intensity value, or a range of values, which separates the objects of interest from the background based

---

6 See [63] for a recent survey.

on their pixel intensities. Despite its simplicity, thresholding can be remarkably effective for certain tasks where there is a clear intensity distinction between regions of interest and the background.

- *Region growing*: Region growing is a pixel-based image segmentation method that starts with a set of seed points and grows regions by adding neighboring pixels with similar properties (such as intensity or texture). The process continues until no more pixels can be added to any region. This technique is particularly useful when the regions of interest are homogenous, and there is a good initial estimate of the seed points.

- *Edge-based segmentation*: Edge-based segmentation methods work by detecting discontinuities in pixel intensities, which often correspond to the boundaries of objects. Edge-based methods often produce fragmented boundaries, especially in the presence of noise or weak edges, and typically require additional postprocessing steps to complete the segmentation.

- *Watershed segmentation*: Watershed segmentation is a technique that treats the image as a topographic surface and finds the "catchment basins" and "watershed lines" in this surface to define regions. It is particularly effective for separating objects in an image that touch each other. However, this technique often suffers from over-segmentation,[7] especially in the presence of noise, and usually requires preprocessing steps or additional constraints to achieve satisfactory results.

---

[7] Over segmentation refers to the condition where an algorithm divides an image into more segments than necessary, often breaking down objects into excessively small parts.

- *Deep learning–based segmentation*: With the advent of deep learning, there has been a significant shift toward using models based on convolutional neural networks for image segmentation, such as U-Net [64], SegNet [65], and many others.[8] These models can learn hierarchical representations from the image data and can capture complex patterns that traditional methods might miss. Despite their success, they require a significant amount of annotated training data and substantial computational resources.

Each of these methods has its strengths and weaknesses, and the choice of segmentation method depends on the specific requirements of the medical image analysis task.

4.4.3 Feature Extraction

Feature extraction is the process of identifying and extracting relevant information from medical images. This information can then be used to classify images, identify abnormalities, or track disease progression. There are numerous feature extraction methods to choose from. They can be classified into the following categories:

- *Intensity-based features*: One of the most basic types of features extracted from medical images is intensity-based features. These can simply be the raw pixel intensities or statistical measures derived from them, such as mean, median, standard deviation, skewness, and kurtosis. Intensity features capture the basic distribution of pixel values within a region of interest. Still, they are highly sensitive to variations in illumination and contrast levels across different images or imaging protocols.

---

[8] See [66], [67], [68], and [69] for recent survey papers on the topic.

- *Texture features*: Texture features capture the spatial distribution and relationship of pixel intensities, measuring the variations, patterns, and regularity in an image. Techniques to extract texture features include gray-level co-occurrence matrices (GLCM), Gabor filters, and local binary patterns (LBPs).[9] Texture features can often capture subtle variations and patterns in the tissue structure that may indicate disease and may not be easily discernable by visual inspection.

- *Shape features*: Shape features, applicable to segmented objects within the image, provide information about the geometric properties of the structures of interest. This can include features such as area, perimeter, compactness, eccentricity, and more complex ones, such as Fourier descriptors. Shape features are often crucial in medical image analysis since many pathological changes manifest as changes in the shape and size of anatomical structures.

- *Local features*: Local feature descriptors, such as Scale-Invariant Feature Transform (SIFT), are algorithms in image processing that identify and describe local features in an image, like corners or blobs. SIFT is particularly known for its robustness to scale, orientation, and lighting changes. Successors to SIFT, such as Speeded Up Robust Features (SURF), Binary Robust Invariant Scalable Keypoints (BRISK), Binary Robust Independent Elementary Features (BRIEF), Oriented FAST and Rotated BRIEF (ORB), and others,[10] have improved upon SIFT's computational efficiency, matching performance, or both, enabling more rapid and

---

[9] See [70] for a recent survey.
[10] See [71] for a recent survey.

accurate local feature extraction and matching in a variety of IPCV applications.

- *Deep learning–based features*: With the advent of deep learning, automatically learned features from raw pixel intensities have shown great promise. CNNs, autoencoders, and other deep-learning architectures can learn hierarchical feature representations that are optimized for the task at hand. These learned features can capture complex, high-level abstractions and have outperformed hand-crafted features on many tasks.

The choice of feature extraction method will depend on the specific requirements of the medical image analysis task and the characteristics of the imaging modality. Often, combining different types of features will yield the best performance.

4.4.4 Prediction

Once features have been extracted, they can be used to make predictions and inferences from the image and derive insights that would not be self-evident from raw pixel values. Prediction tasks usually take the form of *classification* or *regression*.

- *Classification*: In the context of medical image analysis, *classification* refers to the task of categorizing images or regions within images into predefined classes or labels. These classes could represent different tissue types, disease conditions, or anatomical structures. For example, an image of a tumor in a CT scan could be classified as *malignant* or *benign*.
- *Regression*: In the context of medical image analysis, *regression* refers to estimating a continuous

outcome variable, such as the progression of a disease or the likelihood of a patient's recovery based on the patient's medical history and the results of medical imaging tests.

Classification and regression can be performed using various methods, including "classical" machine-learning algorithms and deep-learning techniques.[11]

4.4.5 Post processing

The results from the prediction stage often undergo additional processing steps, including

- *Visual enhancement*: These may include brightness and contrast adjustment, color mapping for better visualization, or the use of false color to highlight certain features to enhance the visual interpretability of the image or the output of the analysis algorithm.

- *Quantitative measurements*: Post-processing often involves extracting quantitative measurements from the output of the analysis. For instance, in a segmentation task, postprocessing might include measuring the volume, surface area, or shape characteristics of the segmented structures. These measurements can provide additional insights beyond what can be visually perceived and can be helpful for the diagnosis, prognosis, or monitoring of diseases.

- *Validation of results*: Validation or verification of the results is another crucial step in post processing. This could involve visual inspection by an expert, comparison with ground truth data (if available), or statistical analysis to assess the reliability and reproducibility of the results. Validation can help

---

[11] See Chapters 3 and 8.

identify errors or anomalies and ensure the results are trustworthy and clinically relevant.

- *Preparation for presentation or storage*: Finally, post-processing often involves preparing the results for presentation or storage. This might involve formatting the results in a way compatible with clinical picture archiving and communication systems or generating reports summarizing the results in an easily digestible format. This step ensures that the results of the medical image analysis can be seamlessly integrated into the clinical workflow.

4.4.6 Visualization and Interpretation

The medical image analysis pipeline culminates in visualization and interpretation, steps that transform complex data into understandable visuals and actionable insights. Visualization techniques help transform the raw data or the output of the analysis into a visually understandable form, allowing healthcare professionals to interpret the images and make clinical decisions. The following are the main visualization techniques in medical image analysis:

- *2D visualization*: This is the simplest form of visualization, where slices of volumetric images are displayed individually. This is the standard form of visualization for modalities like x-ray, where images are inherently 2D. Radiologists are highly trained to read these 2D representations and spot irregularities. Sometimes, a series of 2D images may be animated to simulate a 3D view.

- *3D visualization*: 3D visualization can provide a more comprehensive view of the anatomy, especially for complex structures. This could involve *volume rendering*, where the entire 3D volume is visualized, or *surface rendering*, where only the surface of the structures of interest is displayed. Modern medical image viewers offer interactive 3D visualization, where users can rotate, zoom, and slice the 3D image in real time.

- *Color coding and fusion*: Color coding (also known as *pseudocoloring*) is often used to visualize medical images to highlight certain features. For instance, different tissue types may be colored differently, or different ranges of intensities may be mapped to different colors. In addition, multiple images or image modalities can be fused into a single image for visualization, with different color channels representing other modalities or time points.

- *Quantitative visualization*: Quantitative visualization involves mapping numerical values to visual cues like color or size. This could be used to visualize the output of a classification or segmentation algorithm or to show quantitative measurements extracted from the images. This type of visualization can help in interpreting the results of the analysis and in making data-driven decisions.

- *Advanced visualization techniques* With technological advances, more sophisticated visualization techniques, such as virtual reality or augmented reality for medical imaging, are emerging. These technologies allow for immersive visualization, where the user can interact with the 3D images more intuitively and naturally. While these technologies are still in their early stages of adoption in clinical practice, they hold great potential for improving the interpretation of medical images.

Regardless of the visualization technique used, the user interface and the usability of the software are crucial factors in the interpretation of medical images. The software should be designed in a way that is intuitive and efficient for the user, with features like adjustable window/level settings, easy navigation through the image volumes, and customizable layouts. The usability of the software can significantly affect the user's experience and the efficiency of the interpretation process.

4.5 The Size and Complexity of Medical Image Analysis Using AI

The landscape for AI solutions in medical image analysis is a large and multi dimensional problem space for research and development of innovative solutions.[12] Roughly speaking, we can take into account the following:

- The number of *radiology specialties* (R) – such as abdominal, breast, musculoskeletal, etc.

- The number of *medical imaging modalities* (M) – such as x-ray, CT, MRI, PET, and US

- The number of *anatomical regions* (A) usually associated with each specialty and modality

- The number of *potential findings* (F) associated with a body region

- The number of types of image analysis *tasks* (T) most closely associated with a deep-learning AI

---

[12] This notion of problem space was inspired by talks and slides by Dr. Keith Dreyer, going back to his *Society for Imaging Informatics in Medicine* 2017 Annual Meeting keynote.

solution to assist the human expert in interpreting the associated images – such as classification,
detection, segmentation, etc.

We end up with (at least[13]) $R \times M \times A \times F \times T$ combinations
to consider under the umbrella of "What if we built a deep-
learning solution for *this*?"

For example, *this* could be the *detection* (T) of complete
anterior cruciate ligament tears (F) from the knee (A) MRIs
(M), a task typically associated with musculoskeletal radiology (R).[14] In other words, it addresses one particular point
in this *discrete 5-D space*. Suppose we move ever so slightly
within the space – e.g., only along the F and T axes – and
decide to use deep learning to address the *delineation/seg-
mentation* of *posterior* cruciate ligament tears.[15] We will prob-
ably require different datasets (and associated annotations),
consider other DL architectures, and adopt different metrics
of success (e.g., switching from sensitivity/specificity to Dice
coefficient). In other words, in addition to the problem space
being large,[16] many solutions do not "transfer over" to other
problems – even if they seem to be within a short distance of
one another in this conceptual space.

Let that sink in for a moment. We are looking at potentially *thousands* of meaningful applications of relevant
medical image analysis problems in radiology, many
of which seem ripe for deep-learning AI techniques. It
should not come as a surprise, then, to realize that the
field has exploded in the past decade, making it virtually
impossible to summarize all the relevant developments in

---

[13] This "back of the envelope" calculation does not take into account
equipment settings, selection of cohorts based on several criteria, and
much more.

[14] See [72] for a recent paper on this specific task.

[15] See [73] for a recent paper on this specific task.

[16] Even adopting conservative estimates for R, M, A, F, and T, the number
of meaningful combinations is easily in the thousands.

the literature.[17] Even survey papers[18] have not been able to capture the exponential growth in publications in this space quickly enough, which has led to the publication of meta-reviews[19] summarizing review/survey papers, each of which itself is the summary/review of hundreds of relevant publications in this space.

But wait, there is more.

This vast five-dimensional space might not capture the development of novel applications of deep learning for medical image analysis that did not exist a decade ago, for example, the use of GANs [47], a DL architecture that was invented in 2014, that can be used for generating synthetic imagery, performing denoising in low-dose CT images, and offering alternative approaches for medical image reconstruction, registration, segmentation, detection, and classification [77]. Further emphasizing the astounding pace of advancements in the field, GANs are swiftly being superseded by the more robust, stable diffusion models for image synthesis tasks [78].

The amazingly fast speed of development in the field of AI and deep learning for medical imaging applications can be exhausting. Consequently, in this chapter, we could only scratch the surface of this space. The interested reader will find additional resources for further studies and hands-on explorations in Chapter 10.

---

[17] At the time of writing, a Google Scholar search using "deep learning" "medical imaging" as query returns more than 169,000 results. In contrast, another Google Scholar search using "deep learning" medical image analysis as a query returns 779,000 results!

[18] For another measure of this phenomenon, the survey paper by Litjens et al. [74] – published in 2017 and cited more than 10,000 times since – is the most cited article on artificial intelligence in radiology [75].

[19] See [76] for a recent example of meta-review.

4.6 The Future of Medical Image Analysis

Medical image analysis will continue to be driven by the accelerated development of machine learning and AI. Some trends to keep an eye on include the following:

- *Multimodal and multiscale analysis*: Multimodal analysis [79] combines information from multiple imaging modalities, such as CT, MRI, and PET, to create a more comprehensive view of the patient's condition. This approach allows for better disease characterization, as different modalities can provide complementary information. Multiscale analysis [80], on the other hand, considers information at different scales, from molecular to cellular to tissue levels, to gain a holistic understanding of the disease process. It is increasingly being used in fields such as oncology and neurology.

- *Radiomics and radiogenomics*: Radiomics [81,82] refers to the high-throughput extraction of large amounts of image features from radiographic images, which can be used for quantitative radiology. It can potentially uncover disease characteristics not visible to the naked eye. Radiogenomics [83], a related field, seeks to find the relationship between these image features and genomic patterns, leading to a deeper understanding of the disease and potentially personalized treatment strategies.

- *Time-series analysis and dynamic imaging*: Medical imaging is not always static; many imaging modalities, like functional MRI and dynamic contrast-enhanced CT, produce time-series data that reflect functional or physiological processes. Analyzing these dynamic images can provide valuable insights into disease mechanisms. This area requires specialized analysis techniques that consider the data's temporal aspect.

- *Explainable AI (XAI):* As AI models become increasingly complex, understanding their decision-making process becomes more difficult. This is a significant concern in healthcare, where interpretability and trust are essential. XAI techniques in medical imaging [84] aim to provide additional insight into the predictions made by an AI solution.

- *Edge computing:*   Edge computing refers to the shift of computation closer to the source of data generation, reducing the need for data to be sent to a central location for processing. In medical imaging, this could mean running image analysis algorithms directly on imaging devices or local servers, reducing latency and increasing privacy.

In summary, the amazingly vast and complex field of medical image analysis is constantly evolving, with new AI-based techniques and algorithms being developed at a fast pace. The success of an AI solution for radiology depends on the amount and quality of the data used to train the AI models, a topic we discuss in greater depth in Chapter 5.

Key Takeaways

- Medical image analysis, bolstered by advancements in AI, has revolutionized healthcare by facilitating early disease detection, enabling data-driven decisions, enhancing clinical efficiency, and promoting precision medicine for personalized treatment plans.

- Different imaging techniques such as x-ray, CT, MRI, ultrasound, and PET are used depending on the type of tissue or organ being studied and the disease condition being investigated.

- The image acquisition and reconstruction process associated with each medical imaging modality significantly affects the quality and interpretability of the resulting medical images.

- The medical image analysis pipeline typically consists of *Image Acquisition*, *Preprocessing*, *Image Segmentation*, *Feature Extraction*, *Prediction*, *Postprocessing*, and *Visualization and Interpretation*.

- The realm of AI applications in medical image analysis presents a vast, multidimensional domain ripe for research, with plenty of opportunities for working on innovative solutions to improve accuracy, efficiency, and accessibility.

5

Data: The Essential Ingredient in AI Solutions

5.1 Introduction

Contemporary artificial intelligence (AI) solutions consist of machine-learning (ML)/deep-learning (DL) algorithms that are data driven by nature [85]. The availability of large amounts of data is central to the successful application of AI in radiology: data play a critical role in training, validating, and continuously improving AI models.[1]

The importance of having vast amounts of data, often heralded as the "indispensable fuel" for AI algorithms, is captured perfectly by the now-famous phrase "Data is the new oil" – a concept first publicized by British mathematician and entrepreneur Clive Humby in 2006 [87]. The term draws an analogy between oil and data, likening the role of data in the digital economy to that of oil in the industrial economy. Just as oil, once refined, can be used to power various industrial processes, data can provide valuable insights that drive business strategy and economic growth once data are gathered and analyzed. However, just like crude oil, raw data need processing to be truly valuable and actionable.

---

[1] This is a trend in AI at large, where the recently coined term *Data-Centric AI* is now used to express "an emerging science that studies techniques to improve datasets, which is often the best way to improve performance in practical ML applications" [86].

DOI: 10.1201/9781003110767-5

However, this metaphor is not universally accepted. Some argue that the actual value lies not in the data itself but in the algorithms that transform it – hence the counter phrase, "the algorithm is the gold." In contrast, others interpret Humby's original quote with a touch of irony, suggesting that real-world data can be as messy and challenging to handle as crude oil. While the metaphor underscores the economic potential of data, it also serves as a cautionary reminder of the issues surrounding data privacy, usage rights, and the societal impacts of data misuse – concerns paralleling those associated with oil extraction and use.

In radiology – where data correspond to a complex mixture of medical images, text, and other types of structured and unstructured data – the quality, diversity, and volume of data directly influence the performance of AI algorithms.

In this chapter, we delve deeper into the role of data in AI solutions for radiology. We explore the intricacies of data storage, management, security, and privacy while emphasizing the ethical considerations related to data use in AI. By exploring these facets, we aim to underline the fundamental premise that, when harnessed thoughtfully, data can lead to powerful, efficient, and robust AI solutions in radiology.

5.2 Data Types and Sources in Radiology

5.2.1 Image Data

Medical images, including x-rays, computed tomography (CT) scans, magnetic resonance imaging (MRI), positron emission tomography (PET) scans, and ultrasound (US) images, among others, constitute the primary form of data in radiology.[2] These images, often combined with associated metadata such as patient demographics, medical history, and imaging parameters, serve as the raw material for AI algorithms.

5.2.2 Clinical Data

Clinical data provide a rich backdrop against which AI algorithms can contextualize and interpret medical images. This includes patient demographics, medical history, lab results, and pathology reports. These datasets provide invaluable context and enhance the diagnostic capabilities of AI algorithms in radiology.

5.2.3 Structured Data

Structured data refers to the highly organized information often found in relational databases or spreadsheets. In radiology, this includes patient demographics, lab results, diagnostic codes, procedure codes, and radiology report conclusions. Being easy to search, retrieve, and analyze, structured data provide valuable supplemental information for image analysis.

5.2.4 Unstructured Data

Unstructured data in radiology primarily consist of textual data, such as radiology reports, clinical notes, and referral letters. Although these data are not as neatly organized as structured data, they contain a wealth of information that AI algorithms can leverage through natural language processing (NLP) techniques [88].

---

[2] See Chapter 4 for details.

5.2.5 Other Data Sources

In addition to these primary data types, several additional data sources can further enhance AI solutions in radiology. These include the following:

- Electronic health records (EHRs), which contain a wealth of both structured and unstructured clinical data.

- Genomic data, which provide information about a patient's genetic predispositions and can influence disease prognosis and treatment responses, also holds promise.

- The medical literature at large – constantly updated with the latest research findings – which can be incorporated into AI algorithms to provide the most current evidence-based practice recommendations.

5.3 Data Collection

Radiology, in its traditional form, has relied heavily on the expertise of radiologists who meticulously interpret medical images and annotate findings [81]. Manual annotation, a tedious and time-consuming process, involves labeling medical images to identify and delineate regions of interest or abnormalities [89]. Radiologists' interpretations form the basis of radiology reports, which are crucial sources of clinical information.

However, these traditional methods of data collection are not without limitations. Manual annotation, while invaluable, is prone to inter observer variability and can be subjective [3]. Further, the vast volumes of data generated in radiology departments can be overwhelming, leading to delays and potential inaccuracies in the interpretation process [89].

In response to these challenges, several advancements in data collection techniques have emerged. The advent of digital imaging systems and the adoption of picture archiving and communication systems (PACS) have greatly improved the efficiency of data collection in radiology. PACS store and conveniently retrieve medical images in a digital format, enabling easy access and sharing of imaging data across different clinical settings.

Additionally, the standardization of medical images through the Digital Imaging and Communications in Medicine (DICOM) standards has further revolutionized data collection [90]. DICOM provides a framework for storing, sharing, and printing medical images that also includes patient information, making it an invaluable resource for AI algorithms in radiology.

These advancements have significantly expanded the capabilities and potential of AI in radiology. They allow for the mass digitization, standardization, and easy accessibility of medical images, creating large, diverse datasets for training AI models.

5.4 Data Annotation

Data annotation is a critical step that involves labeling the imaging data to identify specific features or regions of interest, like anatomical structures or lesions. This forms the "ground truth" that the AI algorithms learn from. For instance, in radiology, expert radiologists might manually delineate tumor boundaries on a series of MRI scans.

However, manual annotation is time-consuming and may suffer from inter-rater variability. Semiautomatic and automatic annotation tools are being developed to mitigate these issues, although these still often require expert oversight.

While data annotation is frequently discussed in the context of imaging data in radiology, it is equally applicable and vital to other types of data used in AI applications.

For structured data, which usually come in a predefined format and often include numerical or categorical variables, annotation may involve associating these data points with relevant labels or tags that describe their significance in a medical context – for instance, labeling a series of lab results to indicate the presence or absence of a specific condition.

For unstructured text data, annotation often involves applying NLP techniques to extract valuable information. This might include assigning labels to identify specific medical terms or patient outcomes or categorizing radiology reports based on their findings. This process facilitates the transformation of free-text data into a format that AI algorithms can use.

5.5 Data Preprocessing

The preprocessing stage is an essential part of the typical AI pipeline in radiology, regardless of the data type – be it imaging, textual, or other clinical data. Preprocessing techniques aim to standardize and normalize the data, improving the quality and reducing any bias that may impact the subsequent ML processes.

For image data, preprocessing techniques include image resizing, gray-level normalization, histogram equalization, and noise reduction.[3]

---

[3] See Section 4.4 for details.

For nonimaging data, preprocessing techniques may include data cleaning to handle missing or inconsistent data, data transformation to convert data into suitable forms for analysis, and data reduction to select and retain relevant features.[4] The choice of preprocessing techniques often depends on the type of data, the specific task at hand, and the requirements of the AI algorithm. For instance, text data from radiology reports and EHRs often require NLP techniques to extract useful information. This might involve techniques like tokenization (breaking text into words or phrases), stemming (reducing words to their root form), and removing stop words (common words like "and" or "the" that may not carry significant meaning).

5.6 Data Storage, Governance, and Management

5.6.1 Data Storage Infrastructure

Efficient data storage is a crucial aspect of AI solutions in radiology, as large volumes of imaging and clinical data need to be stored and accessed for analysis. Establishing a robust data storage infrastructure is essential to ensure data availability, integrity, and security.

Traditionally, radiology data were stored in physical film archives, which posed challenges regarding accessibility, space requirements, and long-term preservation. However, with the advent of digital imaging, PACS has become the standard for storing and managing radiological images, offering secure and centralized storage, enabling easy retrieval and sharing of imaging data across healthcare systems [91].

---

[4] See Section 8.2 for details.

Moreover, the increasing adoption of cloud-based storage solutions has provided scalable and cost-effective options for managing large volumes of data. Cloud storage offers advantages such as flexibility, remote accessibility, and automated backups, ensuring data resilience and reducing the burden of maintaining on-premises infrastructure.

5.6.2 Data Governance

Data governance in the context of AI for radiology requires establishing policies, procedures, and guidelines to manage data quality, ensure regulatory compliance, and prioritize patient privacy, informed consent, and data security. Secure practices such as stringent access controls, anonymization techniques, and secure data transfer protocols are vital to protecting patient confidentiality and facilitating data sharing. A comprehensive governance approach enhances data integrity, traceability, and interoperability, fostering transparency and enabling effective collaboration. This, in turn, promotes the responsible use of data, an essential factor in the successful application of AI solutions in radiology.

Furthermore, setting unambiguous guidelines on data ownership, rights, and responsibilities is crucial to cultivating trust and ethical data utilization. Governance frameworks should detail the rules, processes, and accountability structures steering data-sharing practices. Upholding transparency, accountability, and adherence to pertinent privacy regulations like the Health Insurance Portability and Accountability Act (HIPAA) and the General Data Protection Regulation (GDPR) form the backbone of fostering trust among participants in data-sharing initiatives [92].

5.6.3 Metadata Management

Metadata management focuses on capturing and organizing metadata, which provides contextual information about the stored data. This includes details about the data source, acquisition parameters, patient demographics, and any relevant annotations or labels.

Proper metadata management facilitates data discovery, understanding, and utilization, enabling efficient and effective data analysis for AI algorithms.

5.7 Data Quality and Variety

5.7.1 Ensuring Data Quality

Data quality is a critical aspect of AI applications in radiology as it directly influences the accuracy and reliability of the results. Maintaining high-quality data ensures that AI algorithms can produce trustworthy outcomes. Several factors can impact data quality, including noise, artifacts, missing data, and inconsistencies.

To address these challenges, robust quality control measures should be implemented throughout the data collection process. This involves ensuring proper calibration and regular maintenance of imaging equipment to minimize artifacts and distortions, conducting thorough data validation checks, and implementing standardized data acquisition and storage protocols.

Regular monitoring and validation of data quality are vital to identify and rectify any issues that may compromise the accuracy and reliability of AI models.

5.7.2 Dealing with Data Variety

Radiology data exhibit significant variety, both in terms of imaging modalities and the diverse range of clinical information available. This data variety presents unique challenges for AI applications. The integration of different data types, including imaging data, clinical data, and genomic data, requires effective data management strategies.

Data integration techniques [93] play a key role in handling data variety. Harmonizing data formats, standardizing terminologies, and developing data models facilitate interoperability and compatibility between data sources. The integration enables seamless collaboration and analysis across diverse datasets by ensuring consistent representation and organization of data.

Effectively managing data variety in AI for radiology requires a systematic approach to data governance and data management practices (Section 5.6).

5.8 Data Challenges and Considerations

This section focuses on selected challenges associated with the data aspects of AI solutions in radiology: scarcity, heterogeneity, bias, security, and privacy.

5.8.1 Data Scarcity

Data scarcity is a common challenge in AI solutions for radiology [94]. The availability of labeled and annotated data is often limited, particularly for rare diseases or complex conditions. This scarcity of data can hinder the development and training of accurate and robust AI models. Inadequate data quantity may result in overfitting, where the model fails to generalize well to unseen data.

Three possible solutions to overcome data scarcity are data augmentation, synthetic data generation, and data sharing.

Data augmentation involves generating additional training examples by applying transformations, such as rotations, translations, or scaling, to existing data [63].

This can help increase the diversity and variability of the dataset, improving the robustness and performance of AI algorithms.

Synthetic data generation techniques, such as generative adversarial networks (GANs), can also be employed to create artificial data samples that closely resemble real-world data. These synthetic samples can supplement the limited or unbalanced data, enabling better training of AI models [95].

Collaboration and data sharing among healthcare institutions can enable the pooling of data resources and facilitate the creation of more extensive and diverse datasets [96]. Additionally, incentivizing data sharing through privacy-preserving approaches, such as federated learning [97], can address data scarcity while ensuring data privacy and security.

5.8.2 Data Heterogeneity

Data heterogeneity refers to the diversity of data sources, formats, and characteristics in AI applications for radiology. Integrating heterogeneous data poses challenges regarding data preprocessing, normalization, and fusion. Variations in imaging protocols, equipment, and patient populations can introduce biases and inconsistencies, affecting the performance and generalizability of AI models.

Establishing common data formats, ontologies, and metadata standards promotes consistent representation and enables seamless integration of diverse data sources. Data harmonization techniques, such as mapping different terminologies and normalizing imaging protocols, help mitigate the impact of heterogeneity and enhance the comparability and compatibility of data.

5.8.3 Data Bias

Data bias refers to the systematic errors or favoritism in the data used for training AI models. Bias can arise due to various factors, including demographic disparities, variations in data collection practices, or inherent biases in human annotations. Data bias can lead to biased predictions and inequitable outcomes, impacting the fairness and accuracy of AI solutions.

To mitigate data bias, several approaches can be employed [98]. First, efforts should be made to collect diverse and representative datasets that encompass various demographics, clinical presentations, and imaging characteristics. This can help minimize bias stemming from underrepresented groups. Second, incorporating fairness-aware techniques during model development can mitigate bias and ensure equitable decision-making. These techniques involve assessing and correcting for bias in model predictions, promoting fairness and equity across different patient populations.

5.8.4 Data Security and Privacy

Data security and privacy are critical considerations in AI solutions for radiology [96]. Patient data, including sensitive medical images and protected health information (PHI), must be protected from unauthorized access, breaches, and cyber threats.

Adhering to industry-standard encryption protocols and access controls is vital to safeguarding data during storage and transmission. Implementing robust authentication mechanisms, data encryption, and secure network protocols ensures the confidentiality and integrity of patient data [99].

Furthermore, compliance with regulations such as the HIPAA in the United States and the GDPR in the European Union is crucial for maintaining patient privacy and data protection. Strict adherence to these regulations ensures that patient data are handled with the highest standards of confidentiality and privacy [100].

5.9 Interoperability and Standardization

Interoperability and standardization are pivotal components in integrating AI into the medical imaging landscape. Given that medical imaging data often reside in disparate systems within a healthcare facility or across different institutions, effective exchange and interpretation of shared data are crucial for efficient healthcare delivery and advancements in AI.

5.9.1 Standards in Medical Imaging

An essential standard facilitating interoperability in medical imaging is the Digital Imaging and Communications in Medicine (DICOM) standard [90,101], which has provided a universal format for storing and sharing medical imaging data. DICOM ensures consistent and reliable data exchange across various imaging devices and information systems by defining how medical imaging data should be structured and communicated.

Another important standards-developing organization in healthcare data exchange is Health Level Seven International (HL7), which is "dedicated to providing a comprehensive framework and related standards for the exchange, integration, sharing, and retrieval of electronic health information that supports clinical practice and the management, delivery, and evaluation of health services" [102]. Its most recent standard, Fast Healthcare Interoperability Resources (FHIR) [103], allows the representation of clinical and administrative data as interoperable resources. FHIR is designed to be used in various healthcare settings, including EHRs, clinical decision support systems (CDSSs), and medical AI systems. FHIR has demonstrated potential in enhancing interoperability within the broader healthcare context, and it is expected to play a significant role in medical imaging [104].

These are just a few of the most important standards relevant to the interoperability of medical AI systems. The use of standards is essential for ensuring that medical AI systems can communicate with each other and exchange data effectively. This is important for several reasons, including

- It allows for the sharing of medical images and data between healthcare providers and institutions.
- It allows for integrating medical AI systems with other healthcare information technology systems, such as EHRs and CDSSs.
- It allows for developing more powerful and sophisticated medical AI applications.

5.9.2 Challenges in Achieving Interoperability

Achieving interoperability in medical imaging is not without challenges, technical or otherwise [105]. Technical barriers may include variations in data formats, data security concerns, and the complexity of integrating disparate systems and devices. Nontechnical barriers can involve differences in institutional workflows, divergent policies and practices, and the lack of standardized protocols for data exchange. Understanding and addressing these challenges are critical for advancing interoperability in medical imaging and optimizing patient care across healthcare organizations.

Moreover, the variety of standards and their complexity can make ensuring that all systems use the same standards challenging. Healthcare organizations may find it difficult to adopt and use standards effectively.

Despite these challenges, standards and interoperability are essential for the future of medical AI. They can help ensure that systems communicate with each other and exchange data effectively, which will allow for the development of more powerful and sophisticated medical AI applications.

5.10 Ethics of Data Use in AI

5.10.1 Informed Consent for Data Usage

Respecting patient autonomy and ensuring informed consent are essential when using data for AI applications in radiology. Informed consent involves providing patients with comprehensive information about how their data will be used, the potential risks and benefits, and their rights regarding data privacy and sharing. Obtaining informed consent from patients is critical for maintaining transparency and trust and respecting their rights to control the use of their data.

Informed consent processes should be designed to be clear, accessible, and tailored to the specific context of AI research in radiology. This includes providing detailed explanations about the purpose of data usage, the safeguards in place to protect privacy, and the mechanisms for data anonymization and de-identification. Implementing robust informed consent practices ensures that patients are aware of how their data contributes to AI development while protecting their privacy rights [106].

5.10.2 Bias in AI Algorithms

Bias in AI algorithms is a significant ethical concern in radiology. AI algorithms are trained on historical data, which may reflect existing biases and disparities in healthcare. If not appropriately addressed, these biases can perpetuate inequalities and result in biased outcomes, leading to potential harm and unfairness in patient care.

To mitigate bias, it is crucial to ensure diverse and representative training data encompassing various patient demographics, clinical presentations, and imaging characteristics. Rigorous evaluation and validation of AI algorithms should include assessing and correcting for bias and promoting fairness and equity in their predictions and recommendations [107,108].

Ethical responsibility lies with developers, researchers, and healthcare professionals to actively identify and address bias in AI algorithms. Ongoing monitoring, auditing, and transparency in the development and deployment of AI systems can help detect and rectify potential biases, ensuring equitable and unbiased decision-making in radiology.

5.10.3 Potential Consequences of Data Misuse

The potential consequences of data misuse in AI applications highlight the need for robust ethical frameworks and safeguards. Misuse of data can result in privacy breaches, unauthorized access, or unintended consequences, such as erroneous diagnoses or inappropriate treatment recommendations. Ensuring responsible data handling, protection, and adherence to privacy regulations is essential to prevent potential harm.

Implementing comprehensive data governance and security measures, such as encryption, access controls, and regular audits, can help mitigate the risks of data misuse. Data governance frameworks should define the responsibilities and accountabilities of stakeholders involved in AI research and emphasize the importance of upholding ethical standards and patient privacy.

To promote ethical data use, ongoing education and awareness about the potential risks and ethical considerations surrounding AI in radiology are crucial. Healthcare professionals, researchers, and policymakers must engage in continuous dialogue and collaboration to establish guidelines, policies, and regulations that safeguard patient rights and ensure the responsible use of data in AI applications.

In summary, AI's success in radiology relies significantly on the availability of high-quality data. Consequently, adopting widely acceptable standards and best practices for data acquisition, storage, sharing, and management is essential. Furthermore, overcoming challenges related to data integration, bias, privacy, and ethical considerations will drive the responsible and impactful use of data, leading to advancements in AI technologies for radiology.

Key Takeaways

- Data are the essential ingredient in today's AI solutions.
- In radiology, the quality, diversity, and volume of data directly influence the performance of AI algorithms.
- Acquiring, preprocessing, and annotating data for AI applications in radiology can be costly and time consuming.
- There are numerous challenges associated with data acquisition, storage, and sharing. Some are specific to healthcare, such as compliance with HIPAA and GDPR.
- Interoperability plays a fundamental role in medical imaging, allowing data exchange and communication among different systems, devices, and healthcare providers, thanks to standards and protocols such as DICOM and HL7.
- The future of data in AI for radiology will be marked by increasing demand for high-quality data and safe and effective ways of sharing knowledge and advancing state of the art.

6

Clinical Applications
of AI in Radiology

6.1 Introduction

For almost a decade now, radiology has been perceived as a digital image-based specialty that could serve as an early potential testing ground for medical applications of deep learning (DL) solutions for computer vision and image analysis. In Chapter 4, we discussed the frenetic pace of advancements in research projects that claim to have achieved promising solutions in applying DL artificial intelligence (AI) to solve many problems in medical image analysis.

While there has been substantial progress in foundational research in the field of AI for radiology and medical image analysis, a significant gap exists between the successful research results and their clinical implementation as viable products [109,110]. This gap is influenced by several factors, which encompass technical, regulatory, ethical, and cultural challenges.[1]

This chapter presents representative examples of commercially available AI solutions for radiology (*not* research prototypes),[2] organized into two categories: interpretative (Section 6.2) and non interpretative [111] (Section 6.3) uses of AI in radiology.

---

[1] See Chapter 9 for additional discussions.
[2] In some cases, particularly in Section 6.3, in the absence of information about market-ready products, we present representative references to research prototypes for the benefit of the reader.

DOI: 10.1201/9781003110767-6

The information comes primarily from scholarly articles and vendor-supplied product specifications available via two portals:

- The *AI for Radiology: An Implementation Guide* (which emphasizes products available for the European market) [112,113]
- The *ACR Data Science Institute AI Central* (which provides "easy-to-access, detailed information regarding FDA-cleared AI medical products that are related to radiology and other imaging domains") [114]

6.1.1 Overview and Highlights

In the following sections are a few highlights from each portal at the time of writing.[3]

6.1.1.1 *AI for Radiology: An Implementation Guide (European Union [EU])*

- There were 220 products listed on the site, all of which were approved for commercialization in the EU, and about half of which were also cleared by the US Food and Drug Administration (FDA).
- The predominant modality is computed tomography (CT) (with ~40% of the entries), followed by magnetic resonance imaging (MRI) and x-ray.
- More than 35% of the products focus on neuro radiology, followed by chest (~30%), breast, and musculoskeletal (MSK) (~10% of the total number of entries each).

---

[3] This information is dynamic by nature, and the reader is encouraged to check the respective websites for the latest data.

6.1.1.2 American College of Radiology (ACR) Data Science Institute AI Central (US)

- There were 238 FDA-cleared products listed on the site, more than 200 of which have been added between January 2019 and May 2023.

- Out of these, 230 have been cleared for adult use, and only 8 for pediatric use.

- Two-thirds of the products belong to the MIMPS (medical image management and processing system) category.[4] According to the FDA, the software components in this category may provide "advanced or complex image processing functions for image manipulation enhancement, or quantification intended for use in the interpretation and analysis of medical images" [115].

- The remaining one-third are examples of CAD (computer-aided detection and diagnosis), a class of computer systems that aim to assist in the detection and/or diagnosis of diseases through a "second opinion." The goal of CAD systems is to improve the accuracy of radiologists with a reduction of time in the interpretation of images. They are split among the following subcategories:

  - *CADe (computer-aided detection)*: assists in marking/localizing regions that may reveal specific abnormalities.

  - *CADx (computer-aided diagnosis)*: assists in characterizing/assessing diseases, disease types, severity, stage, and progression.

  - *CADe/x (computer-aided detection/diagnosis)*: assists in localizing *and* characterizing conditions.

---

[4] MIMPS is a newly coined term that replaces the use of PACS (picture archiving and communications system) and whose description excludes software functions for the storage and display of medical images.

- *CADt (computer-aided triage)*: assists in prioritizing (triaging) time-sensitive patient detection and diagnosis.
- *CADa/o (computer-aided acquisition/optimization)*: assists in the acquisition/optimization of images/diagnostic signals.
- The predominant modality is CT (with more than half of the entries), followed by MRI and x-ray/mammogram.
- More than 25% of the products focus on the brain, followed by the chest, breast, lung, and heart (~10% of the total number of entries each).

6.2 Interpretative Uses of AI in Radiology

Interpretative uses of AI in radiology typically encompass two main categories: *detection*, where AI aids in identifying abnormalities in radiological images, and *classification*, where AI helps categorize these detected anomalies based on their characteristics.

These are some examples of AI solutions[5] listed in [112] and marked as available for both the European and American markets, organized by subspecialty, as discussed in the following sections.

6.2.1 Abdominal

- [116] provides a workflow for detecting abnormalities on MRI prostate scans.
- [117] is a radiological computer-aided triage and notification software indicated for use in the analysis of

---

[5] The goal of compiling this list is to provide a picture of the breadth and scope of available solutions; company and product names and websites for additional information can be found in the associated reference.

chest and thoracoabdominal CT angiographies. The device is intended to assist hospital networks and trained radiologists in workflow triage by flagging and communicating suspected positive findings of the chest or thoracoabdominal CT angiographies for aortic dissection.

- [118] is a diagnostic aid for liver disorders that uses multiparametric MRI to characterize liver tissue by providing quantitative measures of liver fat and correlates of iron, fibrosis, and inflammation.

- [119] automatically segments the prostate and performs volumetry of the prostate gland.

- [120] is a noninvasive MRI-based solution for measuring the volume fraction of fat in liver tissue.

- [121] is an algorithm that detects and prioritizes intra-abdominal free gas on CT scans with the aim to speed up identification, triage, and treatment.

- [122] transforms any ultrasound device, on-premise, into an AI-powered bladder scanner to measure bladder volume in one click.

- [123] is a semiautomated stone assessment tool designed to aid clinical decision-making by providing clinical metrics about a patient's kidney stone using the patient's noncontrast CT scans.

6.2.2 Breast Imaging

- [124] helps radiologists to identify findings on screening mammograms and uses a color-coded score to provide an at-a-glance level-of-suspicion scoring for screening mammograms.

- [125] provides decision support during the reading of two-dimensional (2D) full-field digital mammography and three-dimensional (3D) digital breast tomosynthesis to help improve clinical accuracy, reduce workload, and optimize workflow.

- [126] calculates and quantifies volumetric breast density as a ratio of fibroglandular tissue and total breast volume estimates and provides these numerical values along with a BI-RADS breast density category to aid healthcare professionals in the assessment of breast tissue composition.

- [127] is a software application designed to assist trained interpreting physicians in analyzing breast and thyroid ultrasound images. It automatically classifies breast lesions suspicious of cancer, based on image data, into one of four ACR BI-RADS or European U1-U5 Classification System–aligned categories. It also categorizes thyroid nodules via the ACR TI-RADS or American Thyroid Association (ATA) risk stratification systems (RSSs), along with a cancer risk assessment.

- [128] is a clinical decision support tool that provides personalized one-, two-, and three-year breast cancer risk estimates based on 2D screening mammograms or 3D breast tomosynthesis.

- [129] is an AI solution that automatically detects suspicious areas for breast cancer on mammograms, including mass, calcification, distortion, and asymmetry.

6.2.3 Cardiac

- [130] is a clinical decision support software for the diagnosis, treatment, prevention, or mitigation of cardiovascular conditions. It allows quantitative analyses of cardiovascular magnetic resonance images.

- [131] is a coronary physiologic simulation software for the clinical quantitative and qualitative analysis of previously acquired CT Digital Imaging and Communications in Medicine (DICOM) data for clinically stable symptomatic patients with coronary artery disease.

- [132] is a web-accessible image post processing analysis software for viewing and quantifying cardiovascular magnetic resonance images.
- [133] is a cardiac magnetic resonance image postprocessing software for the assessment of the heart and greater vessels in both scientific research and clinical routine.

6.2.4 Chest

- [129] is a DL solution that assists radiologists or clinicians in interpreting chest x-rays (posterior to anterior/anterior to posterior). It can automatically detect 10 radiological findings, including atelectasis, calcification, cardiomegaly, consolidation, fibrosis, mediastinal widening, nodule, pleural effusion, pneumoperitoneum, and pneumothorax, and also supports tuberculosis screening. The analysis result contains (i) localization of suspicious areas in color or outline, (ii) abnormality scores reflecting the probability that the detected lesion is abnormal, and (iii) text interpretation for the analysis result by each finding.
- [134] is a radiological computer-aided triage and notification software indicated for use in the analysis of chest and thoracoabdominal CT angiographies. It is intended to assist hospital networks and trained radiologists in workflow triage by flagging and communicating suspected positive findings of chest CT angiographies for pulmonary embolism (PE).
- [135] is a triage and notification software that flags and communicates suspected positive findings of PE and runs automatically on contrast-enhanced CT protocols that include any part of the lungs.

- [136] is an AI clinical decision support solution for chest x-rays, assisting clinicians to interpret chest x-ray studies by detecting up to 124 findings present in emergent, urgent, and nonurgent care settings, including air space opacity, interstitial thickening, volume loss, effusions, and lung masses, pneumothorax, malpositioned lines and tubes, pneumoperitoneum, and acute bony trauma.

- [137] removes ribs and clavicles from a standard chest x-ray, providing a soft tissue image and allowing clinicians to see more. A companion product [138] identifies regions of interest on the bone-suppressed soft tissue image that may be early-stage lung cancer.

- [139] identifies and highlights lines and tubes on portable chest x-ray images.

6.2.5 Musculoskeletal

- [140] uses DL technology for automated measuring of leg geometry to evaluate lower limb deformities, such as leg or lower extremity length discrepancy and knee alignment deformities.

- [141] is an AI solution that detects bone lesions on standard radiographs.

- [142] is a radiation-free software solution that generates a 3D CT-like (synthetic CT) image to visualize bone structures derived from an MRI scan. It can assist in the diagnosis, monitoring and surgical planning, and navigation of orthopedic conditions.

- [143] is an AI companion for lesion detection on bone x-rays. It can detect fractures, effusions, dislocations, and bone lesions.

- [144] is a triage and notification software indicated for use in the analysis of cervical spine CT images.

- [145] is a triage and notification software that flags and communicates suspected positive findings of rib fractures in CT images.

6.2.6 Neuroradiology

- [146] is an AI-based degenerative brain disease diagnosis assistant software that helps clinicians to diagnose brain MRI data through quantitative analysis.
- [147] is a triage and notification software that flags and communicates suspected positive findings of intracranial hemorrhage (ICH) in nonenhanced head CT images.
- [148] provides automated detection and prioritization of acute ICH on noncontrast CT.
- [149] is a cloud-based platform to assist with multiple neuroradiology workflows. It includes third-party modules for brain oncology processing, stroke detection, multiple sclerosis, and neurodegenerative disease diagnostics and follow-up.
- [150] is a triage and notification software that flags and communicates suspected positive findings of brain aneurysms in CT images.

6.2.7 Other

- [151] is a software application designed to assist trained interpreting physicians in analyzing breast and thyroid ultrasound images.
- [152] is an AI assistant software for interpreting lumbar spine MR scans.
- [153] is an algorithm for de-noising CT datasets supporting new CT protocols to drive patient dose down while image quality is maintained.

6.3 Noninterpretative Uses of AI in Radiology

AI plays an increasingly important role in optimizing radiology workflow, enhancing efficiency and productivity. AI-powered applications can automate routine administrative tasks such as data entry and image sorting, allowing radiologists to focus more on interpreting images and providing patient care. These applications play a significant role in today's medical landscape, where there is an increase in workloads and scan complexity, and a need to reduce costs and errors [154].

This section provides an overview of the state of the art on AI applications that have a potential impact on the radiology workflow, organized into four groups: (i) study protocoling, (ii) image acquisition, (iii) worklist prioritization, and (iv) study reporting.[6]

6.3.1 Study Protocoling

Protocoling is the process of choosing the suitable sequences for an MRI or CT scan to ensure that the intended anatomical structures and abnormalities are accurately captured. A radiologist generally carries out this procedure due to their specialized knowledge in the field [154].

One of the most valuable uses of AI is in formulating study protocols and designing hanging protocols suitable for each radiologist. Typically, radiologists develop study protocols based on chosen clinical factors. This process involves examining the clinical data and order information in the patient's electronic health records, referencing relevant lab values, and analyzing previous images and radiology reports. This task is time-consuming – it typically takes

---

[6] See [154] for an expanded and more detailed review.

each sector of a radiology department 1 to 2 hours daily protocoling studies. However, studies have demonstrated that machine learning (ML) algorithms can successfully determine clinical protocols for MSK, brain, and body MRI studies [155].

Furthermore, hanging protocols are crucial for ideal image interpretation, with each radiologist having unique display preferences. Most PACS solutions provide some type of automatic hanging protocol, but this is generally one-size-fits-all, meaning all radiologists view the images displayed similarly. This is gradually evolving, particularly with software solutions that adapt to the user's preferences and arrange the images accordingly. This area is ideally suited for AI application, given that different vendors use various sequence names, and there are inconsistencies in the DICOM metadata, thus preventing the utilization of a straightforward rules-based algorithm for optimizing a hanging protocol [155].

6.3.2 Image Acquisition

Successful interpretation of medical imaging requires proper image acquisition. Factors such as radiation dose, image dimensions, patient movement and positioning, implanted medical devices, and sensor variations can influence the quality of the image being interpreted. Studies have shown that the use of ML techniques in this field can help lower radiation exposure, cut down scan times, decrease the occurrence of false-positive results, and minimize unnecessary repeated imaging, all while preserving image quality [154].

6.3.2.1 Dose Reduction

With the global increase in the use of CT and positron emission tomography (PET) scans, there is a growing concern regarding radiation exposure for patients who frequently undergo these examinations. Radiology departments often face the challenge of balancing between maintaining image quality, managing radiation dose, and adhering to the principle of "as low as reasonably achievable" to minimize unnecessary radiation exposure. The conventional approach to cutting down radiation dose in CT scans involves reducing the tube current, which, while effective, unfortunately, increases noise and diminishes diagnostic confidence [154].

Recent advances in AI techniques for image reconstruction have shown promising results. These techniques can process and improve the quality of low-dose CT images, making them similar in clarity to images acquired with standard doses. DL models can be trained using "noisy" low-quality images and high-quality images, learning to recognize pathologies and normal structures at low doses compared to regular doses. Consequently, the low-dose images can be reconstructed, making them appear as if they were images obtained at regular doses [155].

6.3.2.2 *Image Reconstruction*

Image reconstruction is a fundamental step in medical imaging. Image reconstruction techniques aim at creating high-quality diagnostic images while managing cost, reconstruction time, and risk to the patient. There have been extensive research efforts to use ML techniques to improve image reconstruction in CT, MRI, and PET scans. Examples of targets for improvement include noise reduction, artifact suppression, motion compensation, faster image acquisition, and multimodal image registration. These goals are often interdependent and closely intertwined, thus allowing for the reduction of both radiation dose and contrast agent dose through the successful implementation of image reconstruction techniques [154].

DL–based solutions have been developed to

- De-noise CT and magnetic resonance images without loss of technical detail [156,157].
- Perform MRI noise reduction and artifact suppression (thereby reducing MRI acquisition time) [158–160].
- Reconstruct PET images, originally created using a low-dose radiotracer, making them look as though they were produced using a full-dose radiotracer [161,162].

6.3.2.3 Patient Positioning

Patient positioning in radiology presents several challenges. Firstly, correct positioning is essential for accurate imaging and diagnosis but can be difficult due to factors such as patient discomfort, mobility issues, or lack of cooperation. Second, inconsistent or incorrect positioning can lead to poor image quality, the necessity for repeat scans, increased radiation exposure, and potential delays in diagnosis. Last, specific positions required for certain examinations, such as mammograms, can be physically uncomfortable or cause anxiety for patients, making it challenging to obtain high-quality images.

Recent advances in the use of AI in this space include the following:

- In CT, solutions [163–166] that use a 3D depth-sensing camera that recognizes the anatomical landmarks and models that automatically calculate the patient's center, which is used to optimize the patient bed position for dose and image quality.
- In mammography, solutions [167,168] that automatically evaluate image quality at the time of acquisition to ensure compliance with the *Mammography Quality Standards Act (MQSA) and Program Enhancing Quality Using the Inspection Program [EQUIP]* [169] and/or the

European Reference Organization for Quality Assured Breast Cancer Screening and Diagnostic Services (EUREF) [170] and/or the *Canadian Association of Radiologists Mammography Accreditation Program* [171,172] image quality criteria.

6.3.3 Worklist Prioritization

Traditionally, radiologist worklists are filled with examinations based on predetermined criteria such as the body part to be imaged, the modality of imaging, the patient's location, and the urgency of the examination. However, issues can arise when non-emergency exams are incorrectly ordered as emergencies to hasten the imaging process. This can blur the distinction between regular and emergency cases for radiologists, potentially causing delays in interpreting genuinely urgent cases [154].

In response to this, several AI algorithms have been devised to prioritize examinations with emergent findings across multiple body regions, as referenced in [173]. These models must maintain adequate sensitivity and specificity to detect urgent findings while minimizing the risk of excessive false-positive results.

6.3.4 Study Reporting

Integrating AI applications into radiology reporting can increase the clarity, accuracy, and quality of reporting and decrease report variability in some situations [174]. There has been an increasing level of activity in the use of DL natural language processing models that serve as smart assistants or perform tasks such as automatically populating recommendations for follow-up of incidental findings [175].

In summary, this chapter presented an overview of commercially available solutions in the field of AI in radiology for various tasks in the radiology workflow. This is a fast-moving field, and by the time you read this, the bar will have been raised, and new products and solutions will have appeared in the market.

Key Takeaways

- There are numerous clinical applications of AI in radiology, for both interpretative and noninterpretative tasks.

- There is a growing number of commercially available specialized solutions approved for the European and/or American markets and at least two major public web portals where you can access detailed information.

- Each portal lists 200+ EU-certified and/or FDA-cleared AI products for interpretative radiology across all modalities and subspecialties.

- The availability of solutions for noninterpretative applications of AI in radiology is also growing, with products for crucial workflow tasks, such as study protocoling, image acquisition, worklist prioritization, and study reporting.

7

Harnessing AI in Radiology Education and Training

Michèle Retrouvey, MD
Oge Marques, PhD

7.1 Introduction

The growing role of artificial intelligence (AI) in radiology is revolutionizing the field of imaging. AI algorithms and machine learning (ML) techniques are being increasingly employed to enhance and expedite various aspects of radiology practice, including streamlining daily practice. As experts, radiologists possess the knowledge and skills to ensure the effective use of imaging tools, which include AI. To fully deploy this technology, future radiologists and referring clinicians must have a solid understanding of AI, its limitations, and how to use it to improve patient care. As such, appropriate training for future healthcare professionals is a must. In this chapter, we discuss the integration of AI in radiology education at both undergraduate and graduate levels.

DOI: 10.1201/9781003110767-7

7.2 Importance of Incorporating AI into Medical Education

As AI permeates medicine, trainees should acquire basic AI knowledge, and research suggests that they are eager to do so [176]. Furthermore, AI will play an increasingly prominent role in their future practice, whether they become radiologists or not. AI-powered tools are being used to not only aid image interpretation but also to assist in the appropriate ordering of studies and to help with follow-up. As future experts, medical trainees will need a basic understanding of the tools they use, and as medical educators, it is our job to provide them with a solid foundation that they can build upon.

AI holds immense potential, but it remains imperfect, and as such, it is important to critically evaluate the information generated by these systems. Trainees must master the basics so they can parse through the material presented by these tools and make decisions on whether the algorithms should be trusted or not. With their extensive training, physicians remain the domain expert, and AI systems will always lack the "je ne sais quoi" of a doctor: the ability to think outside the box of humanity [177].

For our learners, incorporating AI into medical education holds immense importance in preparing future healthcare professionals for the evolving landscape of medicine [178]. AI can provide students with access to vast datasets and a wide range of case examples, improving their exposure to diverse pathologies and enhancing their learning experience. AI-powered platforms can simulate real-life scenarios, allowing students to practice interpreting and diagnosing medical images, thereby accelerating their skill development. AI algorithms can serve as valuable decision support tools, assisting radiology students in the interpretation of complex images and detecting subtle abnormalities. By integrating AI into radiology education, students can enhance their diagnostic accuracy, gain confidence, and become more proficient in detecting and analyzing various pathologies. Furthermore, AI can automate time-consuming tasks, such as image segmentation or measurement, enabling radiology trainees to focus on more critical aspects of their training. This improved efficiency can lead to optimized workflow and increased productivity, allowing residents to spend more time learning and refining their diagnostic skills. AI platforms can adapt to individual students' learning needs, providing tailored educational content and feedback. By analyzing students' performance and identifying areas for improvement, AI can help customize educational modules, ensuring personalized learning experiences that address each student's specific learning gaps.

Additionally, trainees equipped with AI knowledge can contribute to the development of AI-driven clinical decision support systems, assisting in personalized treatment plans based on patient-specific data. Understanding the technical aspects of AI allows trainees to critically evaluate and validate AI algorithms, ensuring their appropriate application in patient care. It also enables them to develop highly sought-after skills and opens the door for new opportunities in industry. Ultimately, a comprehensive understanding of AI empowers medical trainees to embrace the future of medicine, contributing to advancements in diagnostics, treatment, and patient outcomes.

7.3 Challenges of Incorporating AI into Medical Education

Incorporating AI into medical education is not without its challenges [179]. Many medical educators do not possess the necessary knowledge and skills to effectively teach AI concepts and applications, and with limited time available in the already packed curriculum, finding space to introduce AI-related content can be difficult. The novelty of AI poses another obstacle, as it may not always be seen as a high priority in medical education. Additionally, the lack of funding earmarked for AI education can impede the development of teaching initiatives, including training resources and infrastructure. Furthermore, there is an inconsistent understanding among educators and learners regarding the role and potential impact of AI in healthcare. This lack of clarity can hinder the integration of AI into the curriculum and result in varying levels of engagement and comprehension [180]. Addressing these challenges requires dedicated efforts to upskill faculty, allocate adequate time and resources, secure funding support, and foster a comprehensive understanding of the role of AI in healthcare among all stakeholders involved in medical education.

7.4 How to Implement AI as an Educator

To successfully implement an AI curriculum at any training level, a structured approach will ensure the target audience's comprehension and practical understanding [181]. Here are some strategies to consider when introducing AI in this context:

1. *Establish the need*: Begin by highlighting the current challenges and complexities involved in image interpretation and analysis, emphasizing the increasing volume of medical imaging data and the need for accurate and efficient diagnosis. Explain how AI can assist in addressing these challenges and improving patient care.

2. *Basics of AI in imaging*: Provide a clear overview of AI concepts as they relate to medical imaging. Explain terms such as ML, deep learning (DL), neural networks, and convolutional neural networks (CNNs). Help students understand how AI algorithms can be trained to analyze images and extract relevant features.

3. *Focus on specific applications*: Narrow down the scope by focusing on specific AI applications in image interpretation and analysis. For example, highlight how AI can aid in detecting and characterizing abnormalities on images, assisting in lesion segmentation, or predicting patient outcomes based on imaging data. Use case examples and real-world scenarios [182] to illustrate the practical applications of AI in these areas.

4. *Demonstrate algorithm outputs*: Demonstrate the outputs of AI algorithms to students by showcasing examples of image analysis using AI software or tools. Highlight the algorithm's ability to identify and highlight regions of interest, provide measurements, or assist in making diagnostic decisions. Encourage students to compare AI-generated outputs with their own interpretations, fostering a better understanding of AI's role as a decision support tool.

5. *Hands-on experience*: Provide opportunities for hands-on experience with AI-powered imaging tools. Utilize interactive sessions or workshops where students can interact with AI software and explore its functionalities. Allow them to analyze sample images using AI algorithms and observe the results. This hands-on experience helps bridge the gap between theory and practice, promoting a deeper understanding of AI applications.

6. *Discuss limitations and potential pitfalls*: Address topics such as algorithm biases, the need for human oversight, false positives/negatives, and the importance of clinical correlation. Students should understand the complementary role of AI as a tool rather than a replacement for clinical judgment.

7. *Continuous learning and updates*: Emphasize the dynamic nature of AI in medical imaging. Encourage students to stay updated with the latest research, advancements, and regulatory considerations related to AI applications. Provide resources such as scientific literature, conferences, or online platforms that focus on AI in medical imaging, ensuring ongoing learning and competence.

By following these strategies, educators can help learners at all levels develop a solid understanding of AI applications in radiology, allowing them to develop the skills to effectively utilize these tools. Now, let's look at incorporating AI education at the undergraduate level in medical school.

7.5 Undergraduate Radiology Education

While the use of diagnostic imaging continues to expand in healthcare, radiology is rapidly changing and adapting to harness the power of AI. Medical students do not require an in-depth understanding of the technical aspects of AI algorithms in radiology, as most medical students will not pursue a career in radiology. However, all medical students will one day order imaging studies and must learn to do so appropriately, particularly as medicine becomes increasingly more reliant on radiology. As tomorrow's healthcare stewards, they must develop a basic understanding of AI's current and future role in radiology, and as healthcare becomes increasingly interdisciplinary, medical students will collaborate with physicians who rely on AI tools in their practice. Understanding the fundamental principles of AI in radiology will enable effective communication and collaboration with radiologists. As a field that relies heavily on written communication, medical students will also interact with radiology reports and order imaging studies in their clinical practice. Understanding how AI is used in radiology will allow them to critically evaluate these reports and the radiologist's recommendations to integrate them into their patient care decisions. Furthermore, the use of AI in medicine raises important ethical considerations, such as patient privacy, algorithm bias, and the role of human oversight. Medical students need a foundational understanding of these ethical issues to navigate the responsible use of AI technologies, regardless of their specialty. Last, AI is a rapidly evolving field, and medical students must embrace the mindset of lifelong learning. Being familiar with the basics of AI in radiology sets the foundation for continuous education and adaptation to future advancements and gives students a solid foundation to build on when future AI technologies are deployed [183].

7.5.1 Integrating AI Basics into the Undergraduate Radiology Curriculum

In today's landscape, it is ideal to integrate radiology vertically and horizontally in the undergraduate medical education (UME) curriculum to ensure comprehensive learning and the development of key competencies. By vertically integrating radiology, concepts and skills are peppered throughout the entire medical curriculum, starting from the preclinical years and continuing into the clinical rotations. Horizontally integrating radiology involves connecting radiology with other disciplines, such as anatomy, pathology, and clinical specialties, to provide a holistic understanding of the role of imaging in patient care. As such, the role of AI in imaging should be both vertically and horizontally integrated [184]. While the lecture series, workshops, and electives available to students may be institution- or course-specific, the framework of the curriculum can be easily adapted or adopted from available programs, including those from the Radiological Society of North America or American College of Radiology. The challenge with existing materials is that they are often geared toward the graduate trainee or board-certified radiologist.

Start with didactic sessions that include basic concepts and terminology related to ML during the preclinical portion of medical school to establish a solid foundation, peppered with practical applications to reinforce concepts. This includes explaining the principles of supervised learning, unsupervised learning, feature extraction, model training, and evaluation metrics. Familiarize trainees with common ML algorithms used in healthcare, such as decision trees, support vector machines, and neural networks. Then, connect the use of ML algorithms to clinical scenarios and emphasize their relevance in medical practice. Illustrate how these algorithms can enhance diagnostic accuracy, predict patient outcomes, or aid in treatment decision-making. This approach helps medical trainees understand the potential benefits of ML algorithms in improving patient care. Showcase real-world examples of successful applications of ML algorithms in healthcare and present case studies or research papers that demonstrate how ML has contributed to improved diagnosis, treatment planning, or patient outcomes. This helps trainees understand the practical implications of ML in radiology.

Hands-on experience is crucial in understanding the practical aspects of AI-powered diagnostic software. Once students have a technical base, they can learn how to navigate software interfaces, interpret AI-generated findings, and integrate them in clinical scenarios. These workshops can include case-based discussions to foster collaboration, and invited experts in AI and radiology can provide feedback and guidance to address questions and concerns and provide insights on how they integrate AI into their daily practice. This, in turn, can lead to collaboration and networking opportunities and discussions on the challenges and limitations associated with AI-powered diagnostic software, including data quality, algorithm validation, potential biases, ethical considerations, and the role of AI in decision-making. In the clinical years, electives may be pursued by interested students to initiate the "physician-as-consultant" role that may be expected of radiologists, a concept explored in Section 7.6.

7.5.2 Evaluating AI Competency in Radiology Undergraduate Medical Education

Evaluating medical students' competency in AI utilization requires a comprehensive approach that assesses their knowledge, skills, and critical thinking abilities in applying AI technologies in healthcare. Here are some methods to consider when evaluating students' competency in AI utilization:

1. *Knowledge assessments*: Use written assessments or multiple-choice questions to evaluate students' theoretical knowledge of AI concepts, including ML, DL, neural networks, and their applications in healthcare. These assessments can cover topics such as algorithm development, ethical considerations, data privacy, and regulatory frameworks. These can be integrated into didactic sessions in the form of audience response activities or can be used in formative or summative examinations.

2. *Case-based scenarios*: Present students with case-based scenarios where they need to apply AI algorithms to interpret and analyze medical images or patient data. Assess their ability to select appropriate algorithms, interpret results, and integrate AI outputs with clinical judgment in making diagnostic or treatment decisions. These can be presented as take-home, individual assignments, or may be completed in groups.

3. *Hands-on practicals*: Conduct practical sessions where students interact with AI software or tools to analyze medical data or images. Assess their proficiency in using AI algorithms and understanding the outputs and their ability to critically evaluate and interpret the results.

4. *Research projects*: Assign research projects or capstone projects that involve the application of AI in a specific medical domain. Evaluate students based on their ability to design, implement, and evaluate AI algorithms, as well as their understanding of the research methodology and ethical considerations involved.

5. *Collaborative assessments*: Promote collaborative learning and assess students' ability to work in interdisciplinary teams. Assign group projects that require students to collaborate with computer science or engineering students to develop and implement AI algorithms for healthcare applications. Assess their teamwork, communication skills, and ability to integrate AI into the healthcare context.

6. *Reflective writing*: Assign reflective writing assignments where students critically reflect on the ethical considerations, potential biases, and challenges associated with AI utilization in healthcare. Evaluate their ability to articulate their understanding of

the ethical implications and their capacity for self-reflection and critical thinking.

7. *Objective Structured Clinical Examination (OSCE)*: Incorporate AI-related stations into OSCEs, where students are assessed on their ability to apply AI algorithms in simulated clinical scenarios. Assess their integration of AI outputs with clinical reasoning, communication skills with patients, and ethical decision-making. Emphasize the importance of their own clinical judgment.

Implement ongoing assessments throughout the curriculum to gauge students' progress and understanding of AI utilization through these techniques, and encourage students to assess their competency in AI utilization through self-reflection exercises or portfolios. This allows them to identify their strengths and areas for improvement, promoting lifelong learning and self-directed development. It is also important as educators to develop objective evaluation metrics and rubrics to ensure consistent and fair assessment of students' competency in AI utilization. Clearly define the criteria for evaluation, and provide explicit guidelines for assessing students' knowledge, skills, and ethical considerations.

7.6 Graduate Radiology Education

Radiology residents will require more AI skills than medical students as they become specialists and shape the field, and they know these skills should be a part of their training [185]. While radiology residents may not be directly involved in developing DL algorithms, understanding their underlying principles and applications is crucial. It enables residents to critically evaluate AI-driven solutions, collaborate effectively with data scientists and technologists, and ensure optimal integration of AI technologies into their daily practice.

The first step to creating a lasting impression is to focus on relevance: How will AI change radiology residents' everyday life? Image analysis, logistics, daily workflow, and research are the four pillars of residency that can be significantly impacted by AI. They are described in more detail next.

7.6.1 Image Analysis

The recent explosion of AI will have deep implications for future practice as these tools move from experimental or research purposes into mainstream healthcare. These may be deployed for image analysis and interpretation in the future, enabling quick, accurate, and consistent information generation about various anatomical structures or pathologies, including measurements that can be used to quantify tumor size, volume, density, or other relevant parameters in an era of precision medicine. Automated methods significantly reduce the time required for these tasks, enabling radiologists to allocate more time to analyzing images, making diagnostic decisions, and interacting with patients and clinicians. Furthermore, automation reduces interobserver variability and enhances the reliability of radiological assessments, which in turn increases quality assurance and validation. As responsible users of AI, trainees need to understand the principles of algorithm validation and must remain vigilant when assessing the accuracy and reliability of automated measurements. Residents will one day be called to protect patients from immature technological advances and must develop strong clinical judgment to do so in the era of AI. In other words, they must learn to trust themselves.

Accurate analysis of medical images is, without a doubt, one of the end goals of current AI technology; this cannot be done without assistance and guidance from experts. As future specialists, radiology residents will benefit from understanding how DL algorithms work and how they can impact radiological image interpretation and potentially capitalize on their knowledge of AI to help further develop the technology as consultants and trainers.

7.6.2 Logistics

The smooth functioning of the entire department is dependent on invisible labor to ensure that patients receive the best care. Trainees may bear the brunt of these tasks or the resulting inefficiencies, and creating a more efficient work environment benefits the entire department and may increase the time available for teaching and learning. So how can AI streamline the day-to-day operations of the radiology department?

1. *Protocol development*: AI algorithms can assist in developing evidence-based protocols for various clinical scenarios. By analyzing vast amounts of patient data and medical literature, AI algorithms can identify patterns, best practices, and optimal treatment pathways. Radiology residents can utilize these AI-driven protocols to help guide how a study will be performed, ensuring personalized and effective patient care.

2. *Workflow analysis and optimization*: AI algorithms can analyze clinical workflows, identifying bottlenecks, inefficiencies, and areas for improvement. By leveraging ML and process optimization techniques, AI can suggest modifications to streamline workflows, enhance resource allocation, and reduce turnaround times.

3. *Predictive analytics and resource planning*: AI algorithms can analyze historical data and real-time

information to predict patient demand, resource requirements, and scheduling needs. This enables proactive resource planning, ensuring sufficient staffing levels, equipment availability, and other necessary resources. Radiology departments can utilize these predictive analytics to optimize resource allocation, minimize wait times, and improve patient access to imaging services. Residents can then be assigned to different services more appropriately.

4. *Quality assurance and error detection*: AI algorithms can be utilized for quality assurance purposes, automatically reviewing and analyzing imaging studies for potential errors or discrepancies. These algorithms can flag abnormalities, ensure adherence to imaging protocols, and identify areas that require further attention. Radiologists can leverage AI-driven quality assurance tools to enhance accuracy, minimize errors, and use these as teachable moments.

5. *Continuous learning and improvement*: AI algorithms can learn from data and feedback, continuously improving their performance and adapting to evolving clinical needs. Radiology departments can utilize AI to track outcomes, monitor performance metrics, and identify opportunities for quality improvement. This iterative process fosters a culture of continuous learning and improvement, enhancing the overall quality of care provided and helping trainees understand their weaknesses.

7.6.3 Daily Workflow

One of the biggest perks of AI is its ability to streamline radiology workflows by automating repetitive and time-consuming tasks to optimize workflow, allowing residents to focus on learning complex cases, reporting critical findings, and delivering quality patient care. By incorporating AI technologies into their training, not only can radiology residents leverage the capabilities of AI algorithms, they can also create a collaborative culture to stay at the forefront of technological advancements in radiology. Let's go through how AI can help at every step of the study, from the initial order to the final communication with the clinical team:

1. *Pre-scan and image acquisition*: AI algorithms can assist in pre-scan and image acquisition processes. AI algorithms can analyze imaging studies and automatically prioritize them based on clinical urgency, ensuring that critical cases are promptly addressed, reducing delays in diagnosis and treatment. Radiologists can focus their attention on the most urgent cases, enhancing patient care and optimizing workflow efficiency. AI algorithms can also help optimize imaging parameters, reducing unnecessary repeat scans and improving image quality.

2. *Image preprocessing and analysis*: AI algorithms can automate image preprocessing tasks, such as noise reduction, artifact correction, and image enhancement. These algorithms can also aid in initial image analysis, highlighting potential abnormalities or regions of interest. Radiology residents can utilize these AI-generated outputs as a starting point for their own detailed analysis and interpretation and gain the expertise to know when to trust the machine and when to trust themselves.

3. *Reporting*: AI algorithms can streamline and optimize workflow by extracting relevant information from the medical record or images and generating structured reports, reducing the administrative burden. By automating routine tasks, residents

can allocate more time to complex case analysis, consultation, and patient interaction.

4. *Clinical decision support*: AI algorithms can provide real-time clinical decision support to radiology residents. These algorithms can integrate patient-specific data, such as clinical history and laboratory results, with imaging findings to offer evidence-based recommendations and treatment suggestions.

5. *Communication*: AI can help radiologists with the important but often tedious task of contacting the ordering physician to report critical findings, thereby improving efficiency and enabling timely patient care.

7.6.4 Research

According to the Accreditation Council for Graduate Medical Education (ACGME) rules for residency training, all residents must participate in scholarly activity during their training. AI can be incorporated into a variety of radiology residency research projects, from image interpretation research to novel algorithm generation.

As AI algorithms require large datasets for training and validation, research projects can focus on feeding algorithms. Images must be carefully curated and organized to ensure data quality and consistency – a process that involves anonymizing patient information, standardizing image formats, and establishing appropriate data storage and management protocols. Once this has been performed, trainees can assist by selecting AI algorithms tailored to their research objectives. Additionally, AI algorithms require training on annotated datasets, where experts label and segment the regions of interest. Residents may become involved in this process, as they can perform image segmentation and help with production of quantitative data.

Once this has been done, residents may be involved in statistical analysis and data interpretation. AI-generated data and measurements can be integrated into statistical analysis to explore correlations, perform hypothesis testing, and derive meaningful conclusions. Quality improvement/quality assurance projects may also be constructed with AI in mind, as described in Section 7.6.2.

Because AI lends itself to collaboration with other domain experts, participation in these types of projects may enable trainees to contribute to interdisciplinary endeavors. Residents can explore the development and application of novel algorithms, participate in algorithm refinement and validation studies, and contribute to the growing field of AI in radiology by interacting with other departments, such as engineering or physics. This involvement fosters a deeper understanding of the algorithms and their potential impact on the field and helps make AI more applicable for healthcare purposes.

7.6.5 Integrating AI into Radiology Graduate Medical Education

As AI will play an increasingly prominent role in supporting clinical practice, the first step to successfully integrating AI into residency training is to establish learning objectives. Clearly defining the specific knowledge areas and skills that residents should acquire given the particularities of their program is the first step, and as AI in training programs becomes more mainstream, new ACGME milestones may emerge to guide program directors, or these new objectives may be listed under an existing milestone. In the meantime, a varied portfolio of AI activities is central to a well-rounded education for radiology residents, which may include didactic lectures but also more hands-on experiences, including case-based learning and interactive workshops [186,187].

7.6.5.1 Case-Based Learning

It is important to create an educational experience that is integrated into real-world care or mimics it as closely as possible. Case-based learning with AI-supported decision-making offers a valuable educational approach for radiology residents, enabling them to use the technology to sharpen their diagnostic skills and improve patient care. It also allows them to further advance their understanding of AI and how it can be used wisely to benefit future generations of radiologists, clinicians, and patients. How do we, as educators, create these experiences? Here are a few tips:

1. *Create realistic clinical scenarios*: Mimic actual patient encounters and integrate AI throughout the entirety of the experience. Consider designing a range of cases that cover different radiological conditions and clinical scenarios and incorporating diverse patient demographics, imaging modalities, and variations in disease presentations. Each case should present relevant clinical information and medical images for analysis. Contrast this AI-enabled scenario with current protocol to help trainees see how incorporating AI alongside traditional workflows can improve the all-around care the patient receives and makes life easier for physicians and staff.

2. *Use AI-assisted image interpretation wisely*: AI algorithms can aid in image interpretation by highlighting potential abnormalities or providing additional insights. Case-based learning can incorporate AI-generated annotations or findings within the radiological images, allowing residents to compare their own interpretations with the AI-generated results. This process encourages critical thinking and facilitates discussions on

the strengths and limitations of AI algorithms in clinical decision-making.

3. *Focus on evidence-based medicine*: Radiologists are consultants, and as such, we must provide evidence-based recommendations based on a combination of patient-specific data, literature, and best practices. Incorporating AI-generated recommendations into case-based learning allows residents to consider AI-generated suggestions alongside their own clinical reasoning. This exposure enhances their ability to make informed decisions by considering both human expertise and AI-supported insights.

4. *Develop a team-based approach*: Case-based learning with AI-supported decision-making encourages collaborative learning among radiology residents, other medical professionals, and AI experts. They can engage in group discussions and interdisciplinary collaborations to analyze cases, interpret AI-generated outputs, and make collective decisions. This collaborative environment fosters a rich exchange of ideas, enhances teamwork, and facilitates learning from each other's perspectives.

It is important to consider the format of these cases. Old-fashioned small group learning may be an easy way to introduce concepts, but if your institution has access to virtual reality headsets or labs, these may provide an immersive and interactive learning experience. Develop interactive features that allow students to manipulate the images, zoom in and out, annotate regions of interest, and interact with the AI-generated outputs. Include decision-making points throughout the virtual cases where trainees interpret the AI-generated findings and make clinical judgments. Prompt them to consider the limitations and potential pitfalls of relying on AI outputs.

7.6.5.2 Interactive Workshops

Creating interactive workshops on AI-powered diagnostic software for residents is an important component for mastery of AI applications in radiological practice. These should be more advanced than those for medical students and focus on understanding the specifics rather than focusing on the general picture. Each workshop should have a clear learning objective: What problem is being addressed by the AI algorithms being presented? Introduce the fundamental concepts and a brief overview of the software and how it performs the task at hand. Demonstrate its features and limitations, then engage the residents either solo or in group activities. Allocate sufficient time for hands-on practice, and be sure to provide guidance and support as needed. At the end of the session, ensure that residents can share their experiences, challenges, and insights gained from the workshop. Consider inviting experts in the field to provide insights and answer students' questions during the workshop.

7.6.6 Developing AI Competency Assessments in Graduate Medical Education

Developing AI competency assessments for radiology residency is an important step to ensure that residents acquire the necessary knowledge and skills and may, in the future, be used to ensure that trainees meet milestones. Designing objective measures to evaluate AI knowledge and skills requires careful consideration of the desired outcomes and competencies and may take multiple forms:

1. *Multiple-choice questions*: Design a set of multiple-choice questions that assess residents' understanding of AI concepts, algorithms, and applications, covering topics such as ML, DL and neural networks, and their relevance to radiology. Ensure that the

questions are focused on radiology practice and that they are clear and concise. The questions may cover a range of difficulty levels to assess different levels of proficiency. These can be administered scattered throughout the 4 years of training at predetermined intervals or all at once, toward the end of training.

2. *Case-based assessments*: Develop case-based assessments that simulate real-world scenarios where residents apply AI knowledge and skills. Provide residents with clinical cases, and ask them to interpret imaging data, utilize AI tools for analysis, and make appropriate clinical decisions based on the AI-generated outputs. Assess their ability to integrate AI into the diagnostic workflow effectively, and include cases where AI is wrong so that residents can learn to trust their own clinical judgment.

3. *Performance evaluation*: Implement objective measures to assess residents' performance in utilizing AI tools and software. For example, evaluate their proficiency in using AI algorithms for image segmentation, measuring anatomical features, or detecting abnormalities. Use standardized metrics to assess accuracy, precision, and recall of their AI-generated results.

4. *Critical analysis and interpretation*: Assess residents' ability to critically analyze and interpret AI-generated outputs. Provide them with AI-generated reports or images, and ask them to evaluate the findings, identify potential limitations or sources of errors, and determine the clinical significance of the AI information. Evaluate their ability to make appropriate decisions based on the AI outputs.

5. *Practical demonstrations*: Require residents to demonstrate their skills in utilizing AI tools and software

through practical demonstrations. This can involve tasks such as performing live demonstrations of using AI algorithms for image analysis, showcasing their ability to customize AI settings for specific clinical scenarios, or presenting case studies where AI has been utilized effectively. This can be done at the workstation or during "hot seat" types of conferences.

6. *Peer review and feedback*: Introduce peer-review sessions where residents evaluate and provide feedback on each other's AI-related work. This can include reviewing AI-generated reports, comparing their own interpretations with the AI outputs, and discussing potential improvements or alternative approaches. Peer feedback can offer valuable insights and foster a collaborative learning environment and is best performed in small groups with a seasoned facilitator.

7. *Continuous learning and improvement*: Encourage residents to engage in ongoing learning and improvement in AI by providing access to educational resources, workshops, and conferences (see Chapter 10 for recommendations). Assess their participation and engagement in these activities to evaluate their commitment to staying updated with the latest advancements in AI.

Remember to align the objective measures with the specific learning objectives and competencies related to AI in radiology. Regularly review and update the assessment methods to reflect the evolving field of AI and ensure that the measures remain valid and reliable.

7.6.7 Challenges of AI Education at the Graduate Medical Education Level

As with undergraduate education, there is a paucity of qualified faculty to teach AI to their residents, and these proficient radiologists are often concentrated at a few institutions. The novelty and lack of funding for the development of AI education are also issues with graduate training in radiology. However, there are additional challenges at more advanced levels as residents are directly responsible for patients in a way that medical students are not. Overuse and misuse of AI may be tempting for busy residents, which may lead to undetected AI errors. Also, residents may be inadvertently uninformed or misinformed of new developments regarding AI products they use routinely, which could have consequences for patient care.

7.6.7.1 Potential Overuse and Misuse

Addressing the limitations of AI in residency training is crucial to ensure its effective and responsible utilization [188]. This includes knowledge of the algorithm's strengths, limitations, and potential sources of errors. By understanding the underlying mechanisms of the algorithm, trainees can better interpret and critically evaluate the AI-generated outputs. Residency programs must educate residents about the constraints and potential biases of AI, emphasizing the importance of maintaining clinical judgment and autonomy. AI should always be used as a tool to enhance decision-making rather than replacing experts. As such, residents should be taught to approach AI-generated outputs with a critical mindset, considering the clinical context, patient-specific factors, and limitations of the algorithms. Regular supervision and feedback from faculty members are essential to monitor the appropriate use of AI and prevent excessive reliance.

7.6.7.2 AI Errors

During the training years, interpretative AI algorithms should be used as a second opinion. They may be helpful as a basis for consulting with colleagues and attendings, reinforcing the understanding that AI is meant to complement their clinical expertise rather than substitute it. By promoting responsible and appropriate utilization of AI, residency programs equip residents with the skills to leverage AI effectively while upholding the principles of patient-centered care. Trainees should also document and communicate any AI-related errors, false positives, or false negatives encountered during their training. Presenting this information to the team in a type of "AI M&M" (morbidity and mortality) conference is valuable not only for algorithm refinement but also as a cautionary tale to other radiologists.

7.6.7.3 Exclusion of Residents

At specific training sites and hospitals, radiology residents should be given regular updates regarding AI products that are used in the department. The evolving performance of these algorithms should also be routinely reported, and regular validation studies and quality assurance processes should be implemented to assess the accuracy and reliability of AI algorithms in different clinical scenarios. Trainees should actively participate in these validation efforts to ensure the proper use of AI and identify any potential sources of errors. This approach ensures that AI is used as a valuable tool in the hands of competent clinicians, enhancing their decision-making capabilities and ultimately improving patient outcomes.

By addressing these challenges, residency training can harness the benefits of AI while maintaining the core principles of patient care and physician expertise. This approach promotes continuous learning, collaboration, and quality improvement, ensuring the responsible and accurate use of AI in radiology practice.

7.7 Ethical Considerations

Ethical considerations in AI adoption are of utmost importance in the integration of AI into radiology education at both the undergraduate and graduate levels. As AI technology continues to advance, there is a need to address ethical implications, including privacy, transparency, bias, and accountability. How should educators approach this delicate topic?

First, introduce medical students to the ethical frameworks that guide the use of AI in healthcare, such as principles of beneficence, nonmaleficence, autonomy, and justice. Discuss how these principles apply to the development, deployment, and evaluation of AI algorithms in medical practice. Teach students about the importance of patient privacy and the responsible handling of healthcare data. Discuss regulations such as the Health Insurance Portability and Accountability Act and General Data Protection Regulation. Emphasize the need for data anonymization, encryption, secure storage, and proper consent processes to protect patient confidentiality.

Alongside privacy is informed consent. Educate residents about the importance of obtaining informed consent from patients when using AI technologies. Discuss the need to explain the purpose, potential benefits, risks, and limitations of AI utilization to patients who may not understand the limitations of what they may consider miraculous technology. Highlight the significance of transparent communication to build trust and ensure patient autonomy and to consider AI-generated outputs as one piece of the diagnostic puzzle, which includes clinical expertise, patient preferences, and individual circumstances.

As educators, it is imperative to touch upon algorithmic bias and its potential impact on healthcare disparities. Biases can be introduced during algorithm development and training processes, and as future radiologists, understanding this inherent flaw is central to understanding the limitations of AI. Engaging trainees in discussions that highlight ethical dilemmas and privacy concerns will help them navigate complex situations involving AI algorithms, patient consent, privacy breaches, or potential biases.

7.8 Future Directions and Potential Impacts

Anticipated advancements in AI technology hold great potential for transforming radiology education. As AI continues to evolve, we can expect advances in areas such as image recognition, natural language processing, and predictive analytics. These developments will enable more accurate and efficient image interpretation, automated report generation, and personalized treatment recommendations. In radiology education, these advancements will shape the curriculum by integrating AI concepts and applications into the learning process. Medical students will gain exposure to AI-powered diagnostic tools, learn how to interpret AI-generated results, and understand the limitations and ethical considerations associated with AI algorithms. By embracing these anticipated advancements, radiology education will empower future healthcare professionals to leverage AI effectively, enhancing diagnostic accuracy, streamlining workflow efficiency, and ultimately improving patient care in the field of radiology.

Collaborative efforts between academia, industry, and regulatory bodies are expected to play a pivotal role in advancing AI technology in radiology education. The synergy between these stakeholders will drive innovation, ensure quality standards, and facilitate the responsible integration of AI into medical training. Academia will contribute by conducting research, developing educational curricula, and fostering a culture of critical thinking and evidence-based practice. Industry partners, including AI developers and medical device manufacturers, will provide expertise, resources, and real-world applications of AI technology. Regulatory bodies will play a vital role in establishing guidelines, standards, and regulations to ensure patient safety, data privacy, and ethical use of AI in healthcare. Through collaboration, these stakeholders can collectively address challenges, share knowledge and best practices, and work toward a shared vision of AI-enabled radiology education. This collaborative ecosystem will drive the development and adoption of cutting-edge AI technologies, empower educators and learners, and ultimately enhance the quality of radiology education and patient care.

Adapting curricula and educational resources to keep pace with AI developments is vital in preparing future healthcare professionals for the evolving landscape of AI in radiology. As AI technology continues to advance, it is essential to integrate AI concepts, applications, and best practices into the educational curriculum. This involves updating existing radiology curricula to include dedicated AI modules or courses that cover the fundamentals of AI, ML, and DL algorithms relevant to radiology. Additionally, educational resources such as textbooks, online platforms, and interactive learning tools should be regularly reviewed and updated to reflect the latest advancements in AI technology. Collaborative efforts between academia, industry, and professional societies can facilitate the development of standardized educational materials that encompass AI-specific knowledge and skills required for radiology practice. Continuous assessment and feedback from educators, trainees, and practicing radiologists can inform the ongoing adaptation of curricula and resources, ensuring they remain relevant, comprehensive, and aligned with the latest AI developments. By continuously adapting curricula and educational resources, radiology education can equip students with the necessary knowledge and competencies to effectively leverage AI technologies in their future practice, ultimately enhancing patient care and outcomes.

In summary, AI algorithms have the capability to revolutionize radiology. By incorporating AI into education, students and residents can gain exposure to cutting-edge technologies and develop a deeper understanding of image interpretation, analysis, and diagnosis. By understanding the basics of AI, medical students can adapt to the changing landscape of healthcare, where AI is becoming increasingly prevalent. Radiology residents can leverage AI tools and technologies to improve patient outcomes, optimize resource utilization, and contribute to the advancement of radiology research.

Key Takeaways

- To ensure effective implementation of AI in radiology, medical students and residents must receive foundational knowledge during their training years.
- AI literacy needs will vary, but a solid foundation will help medical students and radiology residents become lifelong learners and responsible users of AI tools.
- To be an effective AI educator, develop learning objectives, impart tailored AI knowledge, have students and trainees apply it, and have a plan to assess their competence.

8

Getting Started with Deep Learning in Medical Imaging

8.1 Introduction

At first glance, it might seem relatively easy to get started writing code and creating deep-learning (DL) solutions for medical imaging problems, thanks to a myriad of tools, languages, and frameworks; a large number of representative publicly available datasets; numerous offerings in the cloud computing marketplace; and an ever-growing collection of excellent resources available online (books, tutorials, videos, massive open online courses, GitHub repositories, and much more).

The abundance of resources, however, can be overwhelming. There is too much to learn, too many choices to make, and too many details – from utterly irrelevant to mission critical – to consider. Learning artificial intelligence (AI)/machine learning (ML)/DL requires more than just mastering the associated concepts and terminology and developing programming skills using the most popular languages and frameworks. It also requires creating a pathway (or road map) to guide the learning process.

Many pathways and road maps have been proposed for ML in general. In the radiology domain, the work by Wiggins et al. [189] provides an excellent example of how to navigate the journey from foundational knowledge through data curation, coding, and model development, all the way to clinical integration and assessment of the

DOI: 10.1201/9781003110767-8

impact of AI solutions at the point of care. Their companion website [190] offers valuable resources for radiologists (and other medical professionals) to acquire/improve their coding skills in topics related to data science, deep learning, and medical imaging.

This chapter focuses on two main aspects that should help you in your learning journey, regardless of the specific problem you might try to solve and all the associated details (datasets, libraries, neural network architectures, etc.):

1. The generic ML/DL workflow that you should follow (Section 8.2) when building an ML/DL solution.

2. Supplementary practical advice for being successful in the application of DL to medical imaging problems (Section 8.3).

8.2 The Machine-Learning/Deep-Learning Workflow

In this section, we expand upon the brief discussion from Section 3.6 and present advice on how to proceed along the fundamental steps of the ML/DL workflow.[1] There are nine main steps, as follows (see Table 8.1).

8.2.1 Define the Problem and Describe What a Successful Solution Looks Like

As you begin a new project – before starting to collect data, select candidate models, and enter the time-consuming process of several rounds of training your neural networks – take the time to fully understand the problem at hand.

---

[1] They are written in an advisory style, based on the author's experience, primarily in academia, and adapted from [191].

Table 8.1 Summary of the Proposed Workflow

Step	Description Goals	Output
1	Define the problem and success criteria	Clearly define the problem being solved after consulting with subject experts and determining what metrics will be used to evaluate the project.	A clearly documented vision of the problem, desired solution, candidate evaluation metrics, and benchmarks, as well as the context and expectations of the interested constituencies.
2	Acquire data	Collect the appropriate type and amount of data for the task.	Enough high-quality data to enable training, validating, and testing your DL models.
3	Perform exploratory data analysis to gain insights	Clean up, understand, and visualize the data to extract insight on its properties, distribution, range, and overall usefulness.	Solid understanding of the data to be used for your DL project.
4	Prepare the data for your algorithm	Preprocess the data and split the dataset into training, validation, and test sets.	Assurance that your data is formatted in accordance with the input size and format required for the ML model or deep neural network selected for your project.
5	Select a baseline model	Choose an ML algorithm or DL architecture suitable to build a first working solution for the problem, and decide whether to use transfer learning.	An algorithm/architecture/learning paradigm has been decided upon.
6	Train and fine-tune your model	Achieve a desired level of performance in the validation set.	A trained model whose performance on the validation set suggests that it solves the original problem by satisfying an agreed-upon success metric.

(Continued)

Table 8.1 (Continued)

Step	Description Goals	Output
7	Test your model	Evaluate the model's ability to generalize to previously unseen data.	A model whose performance on the test set satisfies an agreed-upon success metric and is ready for deployment.
8	Document and present your solution	Produce rich and descriptive documentation explaining why and how the solution achieves the established objectives.	A rich and accessible presentation that should inform decision-makers to green-light your solution.
9	Deploy and maintain the system	Make the developed solution available to its end users.	A deployed model that meets users' requirements and is systematically tested, monitored, and occasionally updated to reflect the retraining of the model using fresh data.

DL = deep learning; ML = machine learning.

This is the time to connect with the appropriate constituencies and ask questions whose answers should help frame the problem and outline an "ideal" solution.

Once the problem is understood, consider investing time in searching the scientific literature and prior art, understanding existing solutions, how they work, and their limitations.

Next, select a performance measure that is compatible with the project's goals and aligned with the broader business objectives, confirm that subject experts will be available to advise the DL team along the way, write down any assumptions you (or others) have made, and verify those assumptions, if possible.

At the end of this step, you should have a clearly documented vision of the problem, desired solution, candidate evaluation metrics, and benchmarks, as well as the context and expectations of the interested constituencies.

8.2.2 Acquire Data

Once the problem has been clearly specified (and not before), it's time to collect the appropriate type and amount of data for the task. For real-world applications, this step may involve complex legal, financial, and system administration operations (to get authorizations, acquire, and store the data). When designing and running early experiments and proof-of-concept solutions, however, this step may be as simple as selecting a publicly available dataset that might be suitable for the task at hand.

Common mistakes and traps that should be avoided at this stage include the following:

- Not collecting enough data. This is trickier than it appears at first glance because, after all, the answer to the question "How much data do I need for this?" varies significantly between one project and the next, and so does the availability of high-quality data (e.g., images of dogs and cats are much easier to find than specialized medical images depicting rare pathologies).

- Underestimating the important aspect of the *quality* of the data (see Section 5.7). Since most DL solutions operate under the supervised learning paradigm, it is essential that the annotations (e.g., labels) associated with the data samples be as accurate as possible.

- Trusting that acquiring enormous amounts of data by itself will be enough to ensure the project's success. This can backfire in many ways, including the cost of acquiring such data; the additional overhead required to store, clean up, and organize the data; and the unwarranted sense of confidence that "now that we have all this data, all the remaining steps will be trivial."

During this step, you should also ensure that any sensitive information (e.g., patient's protected health information in medical DL solutions) has been de-identified, deleted, or protected somehow. Overlooking these aspects may lead to serious problems in terms of public image, consumer trust, as well as legal and financial consequences.

At the end of this step, you should have acquired enough high-quality data to enable training, validating, and testing your DL models.

8.2.3 Perform Exploratory Data Analysis to Gain Insights

This step, sometimes referred to as *data wrangling*, is fundamentally important to the success of the entire workflow. It involves cleaning up the data, understanding its structure and attributes, studying possible correlations between attributes, and visualizing the data so as to extract human-level insight into its properties, distribution, range, and overall usefulness. Visualization is a fundamental step in exploratory data analysis. In the words of the late Hans Rosling: "The world cannot be understood without numbers. And it cannot be understood with numbers alone" [192].

There is a common misconception in DL: the notion that the deep neural networks' remarkable ability to extract patterns and rules from data eliminates the need for human exploration of the data and its properties. I strongly suggest spending quality time with your data, understanding and trusting it, before moving any further. Invest as much time as you can exploring your data to get a clear sense of what the data at hand warrants in terms of pattern finding and how much the data can be trusted.

At the end of this step, you should be convinced that you understand and trust your data for your DL project.

8.2.4 Prepare the Data for Your Algorithm

Once your data have been acquired, sanitized, and explored manually, it's time to prepare it to be used by your ML/DL algorithm/architecture of choice. The traditional tasks at this step involve setting aside a test set (to be used in step 7 [Section 8.2.7]) and using the rest of the data for training and validation. Note that the details of how to prepare and preprocess your data might depend on your choice of architecture (step 5 [Section 8.2.5]), which reinforces the iterative/cyclical nature of this process.

Moreover, especially if your dataset is small, you should consider using *data augmentation*, a process used to augment the existing dataset in a way that is more cost-effective than further data collection. In the case of medical imaging applications, for example, data augmentation is usually accomplished using simple geometric transformation techniques randomly applied to the original images, such as cropping, rotating, resizing, translating, and flipping.

At the end of this step, you should have defined training, validation, and test sets and ensured that your data are formatted in accordance with the input size and format required by the deep neural network selected for your project.

8.2.5 Select a Baseline Model

This step differs significantly between ML and DL. If you're using conventional ML algorithms, you will typically choose among "classical" ML algorithms (see Section 3.2). If you're considering a DL solution, the choice will be among relatively few (families of) DL architectures for the task at hand (see Section 3.4) and whether the transfer learning paradigm (see Section 3.5) might apply to your problem.[2]

At the end of this step, you should have selected your neural network architecture and decided whether transfer learning can be used to solve the problem at hand.

8.2.6 Train and Fine-Tune Your Model

This is one of the most critical and time-consuming steps in the entire workflow. It consists of training your model and fine-tuning it to achieve a desired level of performance in the validation set. It includes a number of highly technical decisions, such as the choice of optimizer and its hyperparameters (e.g., batch size, learning rate), cost function, number of epochs, and regularization strategies to prevent overfitting, to name just a few. These are rather involved topics about which many DL courses, tutorials, scientific papers, and entire books have been written.

At the end of this step, you should have a trained model whose performance on the validation set suggests that it solves the original problem by satisfying an agreed-upon metric of success.

---

[2] Note that your choice of baseline model in step 5 (Section 8.2.5) might require revisiting step 4 (Section 8.2.4) to ensure that the data have been prepared in a way that is compatible with the selected model.

8.2.7 Test Your Model

Once you're confident (based on the performance evaluation on the validation set) that you have a model that meets the expectations and requirements set forth in step 1 (Section 8.2.1), it is time to evaluate the model's performance on the test set, which will provide a measure of what really matters – the model's ability to generalize to previously unseen data.

At the end of this step, you should have a model whose performance on the test set satisfies an agreed-upon metric of success and is ready for deployment.

8.2.8 Document and Present Your Solution

This is the time to document everything that has been done and present it to the same constituencies in step 1 (Section 8.2.1), explaining why and how the solution achieves the established objectives. I suggest highlighting the big picture first, abstracting complex low-level details, using compelling visuals, and indicating the strengths and limitations of your approach.

At the end of this step, you should have produced a high-level, visually attractive, technically rich (and yet accessible to nonexperts) presentation that should serve as a reference for decision-makers who will eventually green-light your solution.

8.2.9 Deploy and Maintain the System

Once approved, the model is now ready for production. At this stage, you will be involved with many engineering and system integration tasks. Primary concerns will include scalability, systematic testing/monitoring of the model's behavior, and periodic retraining of the model using fresh data. If you're working in a research lab, this step may be optional. For industry projects, it is not only required but significantly more involved and time-consuming than it might appear at first glance [193,194].

8.3 Deep Learning for Medical Imaging: A Recipe for Success

This section provides more specific practical advice tailored to those who want to work at the intersection of AI and medical imaging.

8.3.1 First Things First

Nothing replaces learning by doing or, in this case, learning by coding. As discussed in Section 2.3, the level of insight achieved when writing, running, and modifying code cannot be understated. So, if you haven't done so yet, now is the time to "get your feet wet" by following these steps[3]:

1. Learn the basics of Python – currently the most popular programming language for ML/DL.
2. Learn the basics of a DL framework of choice. Keras, TensorFlow, and PyTorch are good choices.
3. Learn how to use Jupyter notebooks and Google Colab(oratory).
4. Write your first DL solution for a well-known problem, typically by following these steps: (i) using a publicly available dataset; (ii) applying transfer learning using a popular pretrained DL model; (iii) tweaking parameters, hyperparameters, etc., to improve results; and (iv) stopping upon reaching a goal (e.g., good enough accuracy).
5. Share your solution with colleagues (or the world at large), e.g., posting your code on GitHub.

---

[3] See Chapter 10 for learning resources and suggestions.

"Hello World"–type examples [195–197] provide a quick way to get an appreciation for what a working solution for DL in medical imaging looks like.

8.3.2 Ramping Up

If you enjoyed the sensation of accomplishment associated with running the first few code examples, it's time to practice further. The key at this stage is to embark on incrementally harder problems and build your skills and self-confidence along the way. Remember to "walk before you run." For example, "modifying an example of image classification using a certain pre-trained CNN [convoluted neural network] model to use another pre-trained model" should precede "attempting to solve a completely different problem using larger datasets and different DL architectures."

Once you feel ready to tackle more challenging tasks and help advance the state of the art in AI in the radiology space, the advice that follows might come in handy.

8.3.3 Going Further

Here are some recommendations for building successful and impactful DL solutions for medical imaging problems. They should be helpful to researchers who want to make an impact in the field.

This "recipe for success"[4] consists of the following eight steps.

8.3.3.1 Identify Opportunities

Work with intent on what matters. Take the time to define the problem being solved. There are numerous combinations of anatomies, pathologies, image modalities, and

---

[4] The ideas in this subsection appeared in our *Society for Imaging Informatics in Medicine* 2022 abstract "Deep Learning for Medical Imaging: A Recipe for Success" [198], winner of the *Best Poster* award.

tasks (e.g., segmentation, detection, classification) that might still benefit from DL approaches to help solve specialized problems.

8.3.3.2 Search the Literature

Survey papers help get a perspective of the field and maybe narrow it down to a subset of problems of interest. Beware that there has been so much work in this space that there are now systematic meta-reviews of published survey papers (see Section 4.5).

8.3.3.3 Select the Best Tools for the Job

Go beyond general-purpose frameworks, such as TensorFlow or PyTorch. Look also for specialized ones, such as MONAI [199].

8.3.3.4 Build Strong Partnerships

Most problems are large and complex and require multidisciplinary teams with different skill sets – partner with subject matter experts who offer a chance to solve meaningful problems using real data.

8.3.3.5 Enter Challenges and Competitions

Hone your skills and learn what other teams are doing. Hundreds of medical imaging challenges are promoted by leading groups, sites, and professional societies such as the Society for Imaging Informatics in Medicine and the Radiological Society of North America (see Section 2.3). Participating in some of them provides valuable learning opportunities that might not be found elsewhere.[5]

---

[5] Challenges, competitions, and benchmarks have undoubtedly helped advance the state of AI, but they are not without caveats and weaknesses. See [200] for a detailed analysis of challenges in biomedical image analysis and their limitations.

8.3.3.6 Document Code and Data, and Model along the Way

Adopting best practices for dataset and model documentation will help increase transparency, improve communication, and reduce errors at crucial stages in the development life cycle.

8.3.3.7 Publish Your Work

Share your results and findings with the professional community. Be prepared to ensure that your work meets the requirements from leading journals' checklists.

For example, the *Radiological Society of North America's Radiology: Artificial Intelligence* journal encourages use of CLAIM (Checklist for AI in Medical Imaging) [201], a document that has been modeled after the STARD (Standard for Reporting Diagnostic Accuracy) guideline [202] and has been extended to address applications of AI in medical imaging that include classification, image reconstruction, text analysis, and workflow optimization.[6]

8.3.3.8 Go Beyond the Code

Medical imaging is a small part of the whole ecosystem for clinical practice. Once you have achieved successful foundational research results, move further into the path to commercialization.

In summary, building successful AI medical imaging applications requires much more than following the traditional ML/DL workflow and producing promising research prototypes. Medical imaging researchers should also build, document, and test solutions using the best tools, partners, and outlets to disseminate their successes.

---

[6] See also [203] for a higher level look at CLAIM, with justification for each item in the checklist.

Key Takeaways

- It is relatively easy to get started writing code and creating DL solutions for medical imaging problems, thanks to the wide availability of tools, languages, frameworks, publicly available data-sets, cloud computing resources, and learning materials.
- Every ML/DL project should carefully follow a systematic workflow to ensure success.
- Nothing replaces "learning by doing" or, in this case, "learning by coding."
- Mastering coding for DL in medical imaging is a gradual process, going from "Hello World" problems to sophisticated frameworks, large and potentially unwieldy data, and complex model architectures.
- There are several crucial steps to consider before and after coding ML/DL solutions, from identifying opportunities to advance the state of the art on one end, to deploying your work and securing conditions for its commercialization on the other end.

9

The Future of AI in Radiology

9.1 Introduction

The use of artificial intelligence (AI) in radiology is rapidly evolving. AI has the potential to revolutionize radiology by automating tasks, improving diagnostic accuracy, and providing new insights into disease. However, there are also a number of ethical, cultural, and regulatory challenges that need to be addressed before AI can be fully integrated into radiology practice.

This chapter explores three main aspects of the future of AI in radiology:

1. *Ethical and regulatory aspects*: Section 9.2 presents an overview of the ethical considerations of using AI in radiology, such as fairness, bias, privacy, and security. It also discusses how AI is perceived and used in different cultures and how AI is regulated in different countries.

2. *Challenges, controversies, and objections*: Section 9.3 highlights some of the main challenges, controversies, and objections to the use of AI and deep learning (DL) models in radiology. These include adversarial examples, transparency, fairness, bias, explainability (the "black box effect"), data sharing, model sharing, and accountability.

3. *Emerging trends and opportunities*: Section 9.4 discusses the emerging trends in AI for radiology, such as

DOI: 10.1201/9781003110767-9

the rise of large language models (LLMs) [204], the
development of new AI-powered tools, and the integration of multimodal data into AI models.

9.2 Ethical and Regulatory Aspects of the Use of AI in Radiology

The increasing integration of AI in radiology introduces
numerous ethical, cultural, and regulatory challenges.
Ethical issues involve protecting patient privacy, ensuring informed consent, and minimizing algorithmic bias.
Culturally, we must consider the impact of AI on the role
of radiologists and patient interactions while also being
mindful of potential global disparities in AI access and
use. Regulatory challenges vary regionally and continually evolve, requiring stakeholders to keep pace with the
latest developments in the field in order to ensure the safe
and effective use of AI in radiology.

Consequently, it is essential for medical professionals,
regulators, and AI developers to engage in discussions
and collaboratively navigate these challenges to facilitate
the optimal use of AI in radiology. The goal of these discussions should be to develop a framework for the ethical,
cultural, and regulatory use of AI in radiology. In this section, we highlight some ethical and regulatory aspects that
should be kept in mind as such frameworks evolve.

9.2.1 Ethical Considerations

9.2.1.1 Bias and Fairness

One of the most important ethical considerations is the
fairness of AI algorithms. AI algorithms can be biased,
which means that they can produce different results for

different patients depending on their race, ethnicity, gender, or other factors. This can lead to discrimination and injustice, undermining public trust in AI. If patients believe that AI algorithms are biased, then they are less likely to trust the results of AI-powered diagnostic tests.

Addressing bias in AI can be approached from two complementary angles: the data perspective and the model or algorithm perspective. First, the data used should be a true representation of the patient population on which the algorithm will be deployed. This ensures that the data capture the diversity and characteristics of the intended users. Second, the choice of algorithms plays a crucial role. Opting for algorithms specifically engineered to promote fairness[1] can help mitigate the risk of bias, ensuring equitable AI applications.

9.2.1.2 Privacy, Confidentiality, and Security Aspects

Another important ethical consideration is the privacy of patient data [207]. AI algorithms require large amounts of patient data to train, and these data must be protected from unauthorized access.

The potential for unauthorized access to patient data has a number of ethical implications. First, it could lead to the disclosure of sensitive patient information. This could have a negative impact on the patient's privacy and could also lead to identity theft or fraud. Second, the potential for unauthorized access to patient data could lead to the misuse of patient data. For example, patient data could be used to target patients with advertising or to discriminate against patients.

---

[1] There are a number of different fairness metrics that can be used to measure the fairness of an AI algorithm. See [205] and [206] for examples.

There are numerous mechanisms that can be adopted to ensure that patient data are protected, including the use of encryption to protect the data and storing the data in a secure location.

9.2.1.3 The Role of Radiologists

Radiologists could play two key roles in their interaction with AI algorithms in the future.

First, they could utilize AI algorithms as diagnostic aids. AI, when used as a supportive tool, can help radiologists detect potential abnormalities in medical imaging. This could enhance the accuracy of their diagnoses and potentially speed up the process.

Second, radiologists' expertise in anatomy and disease processes can be vital in the development and training of AI algorithms. They could apply their extensive knowledge to work with AI teams to improve the accuracy and efficiency of these algorithms, ensuring they are optimized for use in medical imaging and diagnostics.

9.2.2 Regulatory Considerations

The implementation of AI in radiology, much like in other areas of medicine, is subject to a variety of regulatory bodies to ensure the safe and effective use of this technology. These regulatory bodies provide essential oversight to maintain high standards and provide guidance for the development and application of AI tools.

In the US, the Food and Drug Administration (FDA) oversees the regulation of AI-powered medical devices. The FDA has established specific pathways for the clearance and approval of these devices, such as the De Novo pathway for novel devices of low to moderate risk [208].[2]

---

[2] The regulatory landscape is quickly evolving. See Chapter 10 for additional resources.

In Europe, the European Commission, under the Medical Devices Regulation (MDR) 2017/745 [209] and the In-vitro Diagnostic Devices Regulation (IVDR) 2017/746 [210], oversees the regulation of AI-powered medical devices. The regulations, fully applicable since May 2021, introduce several new aspects, including increased scrutiny of technical documentation, more rigorous clinical evaluation, and post-market surveillance. Furthermore, the European Commission unveiled a proposal in April 2021 for regulating AI systems, which includes specific requirements for high-risk AI systems used in healthcare [211].

It is crucial to note that while these regulatory bodies provide a framework for ensuring the safety and efficacy of AI tools, the rapidly evolving nature of AI technologies presents ongoing challenges for regulatory standards. Constant dialogue between regulators, healthcare professionals, and AI developers is essential to navigate this dynamic landscape.

9.3 Challenges, Controversies, and Objections to the Use of AI and Deep-Learning Models in Radiology

This section briefly highlights some of the challenges, controversies, and objections to using AI and DL models in radiology and gives concrete suggestions for possible solutions.

9.3.1 Technical Challenges

9.3.1.1 Data Quality and Diversity

The efficacy of AI and DL models relies heavily on the quality and diversity of the training data.[3] Unfortunately, access to good-quality, diverse datasets can be a significant

---

[3] See Chapter 5 for a lengthier discussion of this topic.

challenge. Biases in training data can lead to biases in AI system outputs, potentially leading to suboptimal or even harmful patient outcomes. Additionally, the presence of imaging artifacts, inconsistencies in data annotation, and differences in imaging protocols can all compromise data quality.

9.3.1.2 Transparency, Interpretability, and Explainability

AI and DL models, especially those based on complex neural networks, are often referred to as "black boxes" due to their lack of transparency in decision-making processes, which can be a major hurdle in clinical adoption, as healthcare professionals often require a clear understanding of how decisions are made, particularly in critical contexts such as diagnosis and treatment planning [212].

Understanding how AI models make decisions is vital for establishing trust and ensuring the safe and effective use of these tools in a clinical setting.[4] Without this understanding, there's an increased risk of unforeseen errors and biases, which can severely affect healthcare. Therefore, the ability to explain an AI model's predictions – often referred to as *explainability* or *interpretability* – is of great importance.

9.3.1.3 Model Validation and Verification

The clinical use of AI tools requires rigorous validation and verification to ensure their safety, effectiveness, and reliability. However, these processes can be challenging due to the high variability in medical imaging data and clinical scenarios. Furthermore, models that perform well during validation may still fail in real-world applications due to

---

[4] In [213], the authors make a compelling case stating that explainability might not be sufficient to promote trust, and in fact, it might not even be necessary.

the *domain shift* or *model drift,* which refers to the changing data distributions over time or across different settings.

Domain shift [214] can occur when the data used to train an AI model differ from the data the model will use to make predictions. For example, if the patients in the two hospitals have different demographics or imaging protocols, a model trained on chest x-rays from a specific hospital may not perform well on chest x-rays from a different hospital.

Model drift [215] can occur over time as the data distribution changes. For example, a model trained on chest x-rays from 2022 may not perform well on chest x-rays from 2023 if the prevalence of certain diseases changes in the population.

It is important to use a rigorous validation and verification process for AI tools to address these challenges. This process should include the following:

- *Data collection*: The data used to train and validate AI tools should be representative of the data on which the tools will be used to make predictions.

- *Model evaluation*: AI tools should be evaluated on various datasets to ensure they perform well on multiple patient populations and clinical scenarios.

- *Model monitoring*: AI tools should be monitored in real-world use to track their performance and identify potential problems.

9.3.1.4 Data Sharing

Data sharing is the practice of sharing data between different organizations. This can be done for several reasons, including improving the accuracy of AI models, developing new AI models, and researching the effects of AI. Data sharing is essential for the development of AI in radiology because AI models require large amounts of data to train, and it can be challenging to obtain this data from a single organization.

Data sharing in the field of AI, particularly in domains like radiology, presents a set of unique challenges. The primary concerns often revolve around patient privacy, data security, and regulatory compliance.

Medical images and other patient data used in radiology are private and sensitive. While sharing this data to enhance and develop AI algorithms is necessary, it is crucial to do so without breaching patient privacy. Moreover, potential threats like unauthorized access and data breaches add to the data security concerns. The act of sharing data, especially on a large scale, makes it a potential target for cyber attacks.

Another significant issue is regulatory compliance. Laws and regulations, such as the Health Insurance Portability and Accountability Act (HIPAA) in the US or the General Data Protection Regulation (GDPR) in the EU, have strict requirements for the handling and sharing of health data. Adhering to these regulations while sharing data across different regions can be complex and challenging.

Despite these challenges, several possible solutions could facilitate safe and effective data sharing in AI for radiology:

1. *Anonymization and de-identification*: Removing personally identifiable information (PII) from the datasets can help preserve patient privacy. Techniques such as kanonymity or differential privacy [216] can offer robust privacy guarantees while allowing valuable data to be used for AI model development.

2. *Data usage agreements*: Establishing clear data usage agreements can provide legal and ethical

guidelines for data sharing, specifying the terms of use, the responsibilities of all parties, and measures for data security.

3. *Secure multi-party computation*: Techniques like federated learning [97] and secure multi-party computation can enable AI models to learn from data distributed across multiple locations without the need to share the raw data itself.

4. *Synthetic data generation*: Techniques for generating synthetic data, which mimic the properties of real-world data without containing any actual patient information, can be used to create "safe" datasets for AI development and testing.

5. *Blockchain technology*: Blockchain could create a secure, decentralized, and immutable record of data transactions, providing a mechanism for tracking data use and ensuring accountability.

9.3.1.5 Model Sharing

In the context of AI for radiology, *model sharing* refers to making AI models available for other professionals or organizations. This can include open-source models that are publicly available for anyone to use, modify, and improve, or it could involve sharing models within a specific network of professionals or institutions.[5] Model sharing aims to foster collaborative improvements, accelerate research, and reduce duplication of effort, thereby driving faster advancements in AI radiology applications.

Model sharing poses a multifaceted challenge, whose main issues include the following:

---

[5] See [217] for an example of a solution for sharing medical imaging AI models.

1. *Data privacy*: AI models are usually trained on large datasets containing sensitive patient information. Sharing these models might risk exposing these underlying data, especially if the models can be reverse engineered [218]. One potential mitigation strategy involves implementing differential privacy techniques during the model training process. This introduces a level of randomness that helps maintain patient privacy without drastically compromising the model's performance.

2. *Intellectual property rights*: The ownership and rights associated with AI models can be complex, involving the entities that created the model, those that provided the data, and potential users of the model. Clear licensing agreements and contracts should be established that outline usage rights, responsibilities, and any financial arrangements.

3. *Replicability of AI models*: Sharing models is crucial for the replicability and validation of AI research. However, given the variations in data and local clinical practices, a model trained in one institution might not perform as well elsewhere. Besides model sharing, it is also important to share model development and validation methodologies.

Moving forward, to support and enable model sharing, an open and collaborative environment is required, possibly supported by international regulations and guidelines. However, this must be balanced with the necessity to protect patient privacy, respect intellectual property rights, and ensure that models can be safely and effectively applied in different clinical contexts.

9.3.1.6 Accountability

Accountability in any professional field, including the use of AI in radiology, signifies that one can be held answerable for their actions. When it comes to AI models, accountability becomes critically important, given that these models can significantly impact patient health and treatment plans.

These are three means by which AI accountability can be enhanced:

1. *Implementing audit trails*: Audit trails meticulously track the decisions made by AI models, documenting the series of steps leading to a particular decision. This can reveal why the model made a specific choice, shedding light on its underlying logic. It can also provide insight into any potential biases the model might harbor, improving fairness and objectivity.

2. *Increasing transparency*: Transparency involves clear communication about how an AI model works, what data it uses, and how it arrives at its decisions. This disclosure can foster trust among healthcare providers and patients, enabling them to better understand AI outcomes. Greater transparency can also help uncover potential biases within the model, providing an opportunity to rectify them and ensure more equitable AI decisions.

3. *Adopting guidelines for responsible AI use in healthcare, including radiology*: Guidelines (such as [219] and [220]) can outline best practices for developing, training, validating, and implementing AI models, ensuring a robust and ethical AI application.

9.3.2 Controversies and Objections

9.3.2.1 Potential for Job Displacement

The possibility of AI leading to job displacement is a significant area of concern in the realm of radiology (as discussed in Chapter 2). There is a prevailing worry, especially among early-career professionals, that AI could render their roles redundant. While many experts argue that AI will primarily aid radiologists by taking over routine and time-consuming tasks, allowing for a greater focus on complex cases and patient care, the apprehension around job security persists and needs to be addressed [221].

9.3.2.2 Over reliance on AI Systems

Another challenge presented by the integration of AI into radiology is the potential for an overdependence on these systems [222]. If radiologists rely excessively on AI for routine diagnostics, there is a risk of losing proficiency in these areas. This scenario could pose problems if the AI system experiences an error or a technical issue. Furthermore, overreliance could contribute to a de-skilling of professionals, diminishing the critical human element in healthcare.

9.3.2.3 AI's "Black Box" Problem

As noted earlier, the lack of transparency in the way by which many AI and DL models work can lead to distrust and reluctance to adopt these technologies. This is particularly problematic in healthcare, where understanding the decision-making process can be crucial for patient safety and trust.

9.3.3 Addressing the Concerns

9.3.3.1 Potential Solutions to Technical Challenges

Addressing the technical challenges facing AI and DL in radiology requires concerted efforts from researchers, clinicians, and policymakers. Strategies could include developing more robust and transparent models, improving data quality and diversity, and establishing standardized protocols for model validation and verification.

9.3.3.2 Addressing the Human Aspect: Training and Support

Efforts to address the human aspects of AI adoption in radiology should include providing adequate training and support for radiologists, fostering an organizational culture that supports AI adoption, and promoting open and honest discussions about job displacement fears. These efforts include initiatives at multiple levels, from incorporating AI contents into undergraduate and graduate medical education [223,224] (see Chapter 7) to lifelong learning programs for radiologists (see Section 2.3 and Chapter 10).

9.4 Emerging Trends and Opportunities

> It's tough to make predictions, especially about the future.
> –Yogi Berra

Predicting the future is always challenging, even more so in a field that advances as quickly as AI in a field as broad and multifaceted as healthcare.[6]

Following are some of the most prominent emerging trends and opportunities for AI in radiology at the time of writing.

9.4.1 Multimodal Data Integration

AI now facilitates the integration of multimodal data, allowing clinicians to glean more information from a diverse set of patient data [226]. By assimilating radiological images with pathological, genetic, and electronic health record (EHR) data, AI can deliver more precise and

---

[6] For an amusing take on what a Chatbot would say when prompted to discuss the future of AI in medicine, see [225].

personalized diagnostic and treatment insights. The use of machine-learning and DL algorithms for multimodal data fusion [227,228] is an emerging trend that may revolutionize patient care.

9.4.2 Interpretation and Summarization of Medical Records with Large Language Models

Medical image analysis is supported by information stored as text, such as radiology reports and other patient notes. LLMs can be used to analyze reports and other notes [229]. LLMs can also be used to help cope with the ever-increasing research by summarizing new scientific articles and distilling their most essential parts.

9.4.3 AI Applications for Non radiologists

There is a growing number of medical imaging AI applications that are being used by non radiologist clinicians and other healthcare stakeholders. These applications have the potential to improve access to medical imaging and reduce diagnostic errors in low-resource settings and emergency departments, where there is often a lack of around-the-clock radiology coverage. The use of medical imaging AI by non radiologist clinicians is still in its infancy, but it has the potential to significantly improve access to medical imaging and patient outcomes [230].

9.4.4 Generalist Medical AI Models for Radiology

A new generation of generalist medical AI models that can perform the entire task of radiological image interpretation is on the horizon. These models will be able to accurately generate a full radiology report by interpreting a wide range of findings in images, including the degree of uncertainty and specificity of each finding.

They will also be able to fuse clinical context with imaging data and leverage previous imaging in their decision-making. Rapid developments in AI models, including self-supervised models, multimodal models, foundation models [231], and LLMs for text data and combined image and text data, have the potential to accelerate progress in this area [230].

In summary, the use of AI in radiology raises a number of challenges that need to be addressed before AI can be fully integrated into radiology practice.

It is important to have a thoughtful and informed discussion about the ethical, technical, and regulatory aspects of AI in radiology. This discussion should involve radiologists, AI experts, healthcare providers, patients, policymakers, and the public.

The goal of this discussion should be to develop a framework[7] for the ethical, cultural, and regulatory use of AI in radiology. This framework will help to ensure that AI is used in a way that is safe, effective, and ethical.

Key Takeaways

- AI has the potential to revolutionize radiology, but there are numerous challenges that need to be addressed before AI can be fully integrated into radiology practice.

- Successful deployment of AI solutions in radiology must ensure that ethical considerations, such

---

[7] Table 1 in [232] provides a general framework for considering short-, medium-, and long-term clinical AI quality and safety issues in medicine.

as fairness, bias, privacy, confidentiality, and security, are followed.

- The adoption of AI in radiology is also subject to ever-evolving regulatory requirements.

- Some of the technical challenges to be overcome to ensure the successful use of AI in radiology include data quality and diversity, the opaqueness of DL models, domain shift, model drift, data and model sharing, and accountability aspects.

- There are still controversies and objections to the use of AI in radiology, including job displacement, the risk of over reliance on decisions made by AI systems, and the limited trustworthiness afforded by some "black box" solutions.

- There is a potentially bright future ahead, with many emerging trends and opportunities fueled by the latest advancement in AI, such as foundation models and LLMs for the combined processing of image and textual data.

10

Resources for Further Learning

This chapter contains pointers to curated resources that you might want to explore to learn more about the latest developments and best practices in artificial intelligence (AI) and deep learning (DL), as well as the social, moral, ethical, and philosophical implications of DL and AI in radiology and healthcare at large. They are organized in a way that roughly maps to the chapter where they might be most relevant. Selected references and resources appear with a star (★) to indicate their relevance. Some references might include a brief comment or summary. We hope that the suggested books, online courses, scientific articles, white papers, videos, and other resources will set the tone for a successful lifelong learning journey.

Chapter 1

There are a great many books, journals, and online resources where you can learn more about AI and its impact on healthcare and medicine. Here are some carefully selected suggestions.

Books

- ★ S. Russell and P. Norvig, *Artificial Intelligence: A Modern Approach*. Pearson, 4th ed., 2020 [4]
  - *This is a widely known reference textbook in AI, last updated in 2020 and adopted by more than 1,500 colleges and universities worldwide. Additional resources at:* https://aima.cs.berkeley.edu/

DOI: 10.1201/9781003110767-10

- ★ E. Topol, *Deep Medicine: How Artificial Intelligence Can Make Healthcare Human Again*. Hachette UK, 2019 [233]
  - *All-around excellent book by a thought leader at the intersection of AI and medicine.*
- ★ K. L. Holley and S. Becker, *AI-First Healthcare*. O'Reilly Media, Inc., 2021 [234]
  - *Highly readable book with emphasis on providing guidance to healthcare informatics leadership teams on how to create AI strategy and implementation plans for healthcare.*
- B. Lorica, *What Is Artificial Intelligence*. O'Reilly Media, Inc, 2016 [7]

Scientific Articles

- ★ P. Rajpurkar, E. Chen, O. Banerjee, and E. J. Topol, "AI in health and medicine," *Nature Medicine*, vol. 28, no. 1, pp. 31–38, 2022 [235]
- M. Mitchell, "Why AI Is Harder than We Think." https://arxiv.org/abs/2104.12871, 2021 [236]
- ★ B. Meskó and M. Görög, "A short guide for medical professionals in the era of artificial intelligence," *NPJ Digital Medicine*, vol. 3, no. 1, 2020 [237]
- ★ E. J. Topol, "High-performance medicine: The convergence of human and artificial intelligence," *Nature Medicine*, vol. 25, pp. 44–56, Jan. 2019 [5]
- A. Esteva, A. Robicquet, B. Ramsundar, V. Kuleshov, M. DePristo, K. Chou, C. Cui, G. Corrado, S. Thrun, and J. Dean, "A guide to deep learning in healthcare," *Nature Medicine*, vol. 25, no. 1, pp. 24–29, 2019 [238]
- S. Mukherjee, "A.I. versus M.D." https://www.newyorker.com/magazine/2017/04/03/ai-versus-md, March 2017 [239]

Courses

There are many online courses and boot camps to choose from, offered by reputable universities and companies, often available on major massive open online course (MOOC) platforms, such as Coursera, Udacity, or edX. For absolute beginners in AI, I personally recommend Andrew Ng's "AI for Everyone" [240].

Chapter 2

In addition to the dozens of references cited in Chapter 2, here are some suggestions for scientific articles to enhance your appreciation of the interplay between AI and radiology and the impact on the radiologist's role.

- P. Rajpurkar and M. P. Lungren, "The current and future state of AI interpretation of medical images," *New England Journal of Medicine*, vol. 388, no. 21, pp. 1981–1990, 2023 [230]

- J. Banja, "AI hype and radiology: a plea for realism and accuracy," *Radiology: Artificial Intelligence*, vol. 2, no. 4, 2020 [241]

- F. Pesapane, M. Codari, and F. Sardanelli, "Artificial intelligence in medical imaging: Threat or opportunity? Radiologists again at the forefront of innovation in medicine," *European Radiology Experimental*, vol. 2, pp. 1–10, 2018 [242]

- A. Hosny, C. Parmar, J. Quackenbush, L. H. Schwartz, and H. J. Aerts, "Artificial intelligence in radiology," *Nature Reviews Cancer*, vol. 18, no. 8, pp. 500–510, 2018 [89]

Chapter 3

There are countless books, journals, and online resources where you can learn more about machine learning (ML) and DL. The curated list of ML books by Jason Brownlee

[243] is an excellent starting point. You should also check this list of ML and DL frameworks, libraries, books, blogs, courses, and software on GitHub [244].

Following are some additional recommendations.

Books

- ★ A. Geron, *Hands-On Machine Learning with Scikit-Learn, Keras, and TensorFlow*. O'Reilly Media, Inc., 3rd ed., 2022 [59]
  - *This is one of my favorite books on ML, DL, and related topics. Companion examples at:* https://github.com/ageron/handson-ml3

- ★ A. Ng, *Machine Learning Yearning: Technical Strategy for AI Engineers in the Era of Deep Learning.* 2019 [245]
  - *This short book by one of the most recognizable names in AI distills years of experience and provides practical advice.* https://www.mylearning.org

- ★ M. Nielsen, *Neural Networks and Deep learning.* http://neuralnetworksanddeeplearning.com [246]
  - *This free book – available exclusively online – contains excellent explanations of the fundamentals of neural networks and deep learning. Companion code at:* https://github.com/mnielsen/neural-networks-and-deep-learning

- I. Goodfellow, Y. Bengio, and A. Courville, *Deep Learning.* MIT Press, 2016 [247]

Courses

There are innumerable online courses and boot camps to choose from, offered by reputable universities and companies, often available on major MOOC platforms. I personally recommend the foundational courses from the *Google Machine Learning Education* portal [248].

Other Resources

- ★ For a broad understanding of ML and related topics, I strongly recommend Cassie Kozyrkov's *Making Friends with Machine Learning* course on YouTube [249].

- ★ The videos from the *3Blue1Brown* YouTube channel explain neural networks in a technically sound and visually pleasant way [250].

- This explanation of convoluted neural networks (CNNs) [251] from Stanford's landmark *Deep Learning for Computer Vision* course can be very helpful.

- Chris Olah's blog contains richly illustrated explanations of topics related to neural networks, including CNNs [252] and long short-term memory networks [253].

Chapter 4

Books

Medical image processing and analysis are deep and complex topics with numerous textbooks to choose from. Here are some recommendations:

- P. Suetens, *Fundamentals of Medical Imaging*. Cambridge University Press, 3rd ed., 2017 [61]

- A. P. Dhawan, *Medical Image Analysis*. John Wiley & Sons, 2nd ed., 2014 [60]

- W. Birkfellner, *Applied Medical Image Processing: A Basic Course*. CRC Press, an imprint of Taylor and Francis, 2nd ed., 2014 [254]

Additionally, there is a vast literature on radiographic image analysis and richly illustrated volumes on visual approaches to diagnostic imaging, including

- K. M. Martensen, *Radiographic Image Analysis.* Elsevier Health Sciences, 2013 [255]
- J. Mandell, *Core Radiology.* Cambridge University Press, 2013 [256]

On the interface between AI and medical imaging, my favorite book is

- ★ E. R. Ranschaert, S. Morozov, and P. R. Algra, *Artificial Intelligence in Medical Imaging: Opportunities, Applications and Risks.* Springer, 2019 [257]

Scientific Articles

- A. Esteva, K. Chou, S. Yeung, N. Naik, A. Madani, A. Mottaghi, Y. Liu, E. Topol, J. Dean, and R. Socher, "Deep learning-enabled medical computer vision," *NPJ Digital Medicine*, vol. 4, no. 1, p. 5, 2021 [258]

Tools and Programming Resources

During the past two decades, thanks to the popularization of libraries, frameworks, and toolboxes, the building blocks of the medical image analysis pipeline have become commodities.

Popular tools, packages, and libraries include

- ImageJ [259] and Fiji [260]
- ★ MATLAB [261] (and its toolboxes[1])
- OpenCV [262]
- Pillow (Python Image Library fork) [263]

---

[1] Notably the Image Processing, Deep Learning, Medical Imaging, and Computer Vision toolboxes.

Chapter 5

Books

- ★ A. Nguyen, *Hands-On Healthcare Data*. O'Reilly Media, Inc., 2022 [264]
  - *Recently published book that teaches how to implement solutions for working with healthcare data, from data extraction to cleaning and harmonization to feature engineering. Companion code at:* https://gitlab.com/hands-on-healthcare-data

Scientific Articles

- O. Diaz, K. Kushibar, R. Osuala, A. Linardos, L. Garrucho, L. Igual, P. Radeva, F. Prior, P. Gkontra, and K. Lekadir, "Data preparation for artificial intelligence in medical imaging: A comprehensive guide to open-access platforms and tools," *Physica medica*, vol. 83, pp. 25–37, 2021 [265]
- ★ M. J. Willemink, W. A. Koszek, C. Hardell, J. Wu, D. Fleischmann, H. Harvey, L. R. Folio, R. M. Summers, D. L. Rubin, and M. P. Lungren, "Preparing medical imaging data for machine learning," *Radiology*, vol. 295, no. 1, pp. 4–15, 2020 [266]

Other Resources

- The MONAI Label End-To-End Tutorial Series [267] (described in [268]) offers a practical example of what it takes to label medical images.

Chapter 6

Scientific Articles

Since Chapter 6 focuses primarily on commercially available AI solutions (that once were research prototypes), we deemed it important to suggest these two articles from a landmark *National Institutes of Health/Radiological Society of North America/American College of Radiology/The Academy* workshop in 2018 since they provide road maps for foundational and translational research on AI in medical imaging:

- C. P. Langlotz, B. Allen, B. J. Erickson, J. Kalpathy-Cramer, K. Bigelow, T. S. Cook, A. E. Flanders, M. P. Lungren, D. S. Mendelson, J. D. Rudie, *et al.*, "A roadmap for foundational research on artificial intelligence in medical imaging: From the 2018 NIH/RSNA/ACR/The Academy workshop," *Radiology*, vol. 291, no. 3, pp. 781– 791, 2019 [269]
- B. Allen Jr, S. E. Seltzer, C. P. Langlotz, K. P. Dreyer, R. M. Summers, N. Petrick, D. Marinac-Dabic, M. Cruz, T. K. Alkasab, R. J. Hanisch, *et al.*, "A road map for translational research on artificial intelligence in medical imaging: From the 2018 NIH/RSNA/ACR/The Academy workshop," *Journal of the American College of Radiology*, vol. 16, no. 9, pp. 1179–1189, 2019 [270]

Additionally, the articles that follow provide insights into the growing market for US Food and Drug Administration (FDA)-cleared AI products for medical imaging:

- M. Khunte, A. Chae, R. Wang, R. Jain, Y. Sun, J. Sollee, Z. Jiao, and H. Bai, "Trends in clinical validation and usage of US Food and Drug Administration-cleared artificial intelligence algorithms for medical imaging," *Clinical Radiology*, vol. 78, no. 2, pp. 123–129, 2023 [271]
- M. Milam and C. Koo, "The current status and future of FDA-approved artificial intelligence tools in chest radiology in the United States," *Clinical Radiology*, vol. 78, no. 2, pp. 115–122, 2023 [272]
- D. P. Stonko and C. W. Hicks, "Mature AI/ ML-enabled medical tools impacting vascular

surgical care: A scoping review of late-stage, FDA approved/cleared technologies relevant to vascular surgeons," in *Seminars in Vascular Surgery*, Elsevier, 2023 [273]

Other Resources

- The official FDA page for documents and updates on topics related to "Artificial Intelligence and Machine Learning in Software as a Medical Device" [274] is worth bookmarking.

Chapter 7

While curriculum efforts in undergraduate and graduate medical education are starting to take shape, medical students, radiologists, and medical imaging informatics professionals can take advantage of educational initiatives, such as the following:

- The Data Science Pathway (DSP), a pilot program for fourth-year residents at Duke University, developed by Walter F. Wiggins, MD, PhD, in collaboration with the MGH & BWH Center for Clinical Data Science (CCDS) [189], whose companion website offers valuable resources for radiologists (and other medical professionals) to acquire/improve their coding skills in topics related to data science, DL, and medical imaging [190].
- Society for Imaging Informatics in Medicine (SIIM) Hackathons [275], which have created a lively atmosphere of friendly competition during every SIIM Annual Meeting, and whose resources (servers, datasets, libraries, APIs, etc.) remain available 24/7/365.
- On-demand courses, such as the ones available at the Radiological Society of North America (RSNA)

Learning Center [276], instructional videos and materials and interactive no-code examples of DL/ AI in radiology in action in the AI-LAB [277], an initiative of the Data Science Institute (DSI) of the American College of Radiology (ACR).

- The *National Imaging Informatics Curriculum and Course – Radiology (NIIC-RAD)* [278], a partnership between RSNA and SIIM, a week-long, online course that introduces the fundamentals of imaging informatics and includes topics related to AI – and how it will *not* replace radiologists.

- Several certificate programs and courses by leading universities and MOOC platforms, including the Stanford/Coursera *Fundamentals of Machine Learning for Healthcare* course [279], part of the *AI in Healthcare Specialization*.

- The RSNA *Imaging AI Certificate Program* [280], which – at the time of writing – offers two options:

  - A *Foundational Certificate* course that teaches how to confidently evaluate, deploy, monitor, and use AI tools

  - An *Advanced Certificate* course that focuses on gaining a deeper understanding of how to deploy, monitor, and use AI solutions in a clinical setting

Chapter 8

The suggestions that follow focus primarily on Section 8.3 and cover the scope ranging from learning the basics of Python to publishing your work and going beyond the code.

Books

- ★ F. Chollet, *Deep Learning with Python*. Shelter Island: Manning Publications, 2nd ed., 2021 [58]

- *This is an excellent book to learn the basics of DL with Keras, written by Keras' creator and DL researcher, François Chollet.*
- J. Howard and S. Gugger, *Deep Learning for Coders with fastai and PyTorch*. O'Reilly Media, 2020 [281]

Kaggle Courses

These (free, interactive, self-paced) courses – taught in the Kaggle Notebook self-contained programming environment – provide a solid practical foundation for beginners.

- Python [282]
- Intro to Machine Learning [283]
- Intermediate Machine Learning [284]
- Intro to Deep Learning [285]

Other Resources

- Awesome Python [286]
  - *A curated list of Python frameworks, libraries, software, and resources*
- Git Tutorial [287]
  - *A good tutorial to learn/refresh basic knowledge of Git and GitHub*
- Official PyTorch tutorials [288]
- Official TensorFlow courses and tutorials [289]
- Python for Non-Programmers [290]
  - *An extensive list of interactive tutorials and other online resources for learning the basics of programming in Python*

Chapter 9

Scientific Articles

- T. Boeken, J. Feydy, A. Lecler, P. Soyer, A. Feydy, M. Barat, and L. Duron, "Artificial intelligence in diagnostic and interventional radiology: Where are we now?" *Diagnostic and Interventional Imaging*, vol. 104, no. 1, pp. 1–5, 2022 [291]

- C. Garbin and O. Marques, "Assessing methods and tools to improve reporting, increase transparency, and reduce failures in machine learning applications in health care," *Radiology: Artificial Intelligence*, vol. 4, no. 2, p. e210127, 2022 [292]

- E. Wu, K. Wu, R. Daneshjou, D. Ouyang, D. E. Ho, and J. Zou, "How medical AI devices are evaluated: Limitations and recommendations from an analysis of FDA approvals," *Nature Medicine*, vol. 27, no. 4, pp. 582–584, 2021 [293]

In summary, this chapter concluded our coverage of the fascinating field of AI in radiology by offering you carefully curated resources to inspire your lifelong journey in this field. I hope this book has prepared you for the next steps and offered valuable resources to help you shape your own learning path.

Good luck and happy learning!

References

[1] D. Silver, T. Hubert, J. Schrittwieser, I. Antonoglou, M. Lai, A. Guez, M. Lanctot, L. Sifre, D. Kumaran, T. Graepel, T. Lillicrap, K. Simonyan, and D. Hassabis, "A general reinforcement learning algorithm that masters chess, shogi, and go through self-play," *Science*, vol. 362, no. 6419, pp. 1140–1144, 2018.

[2] L. Weidener, M. Fischer, *et al.*, "Artificial intelligence teaching as part of medical education: Qualitative analysis of expert interviews," *JMIR Medical Education*, vol. 9, no. 1, p. e46428, 2023.

[3] J. H. Thrall, X. Li, Q. Li, C. Cruz, S. Do, K. Dreyer, and J. Brink, "Artificial intelligence and machine learning in radiology: Opportunities, challenges, pitfalls, and criteria for success," *Journal of the American College of Radiology*, vol. 15, no. 3, Part B, pp. 504–508, 2018.

[4] S. Russell and P. Norvig, *Artificial Intelligence: A Modern Approach*, 4th ed. Upper Saddle River, NJ: Pearson, 2020.

[5] E. J. Topol, "High-performance medicine: The convergence of human and artificial intelligence," *Nature Medicine*, vol. 25, pp. 44–56, Jan. 2019.

[6] K. D. Foote, "A brief history of artificial intelligence," Apr. 2016. Available: www.dataversity.net/brief-history-artificial-intelligence/ [Accessed June 15, 2023].

[7] B. Lorica, *What Is Artificial Intelligence*. Sebastopol, California: O'Reilly Media, Inc., 2016.

[8] D. Silver, T. Hubert, J. Schrittwieser, I. Antonoglou, M. Lai, A. Guez, M. Lanctot, L. Sifre, D. Kumaran, T. Graepel, T. Lillicrap, K. Simonyan, and D. Hassabis, "Mastering chess and shogi by self-play with a general reinforcement learning algorithm," 2017. Available: https://arxiv.org/abs/1712.01815.

[9] E. Strickland and G. Zorpette, "The AI apocalypse: A scorecard – IEEE spectrum," June 2023. Available: https://spectrum.ieee.org/artificial-general-intelligence [Accessed June 28, 2023].

[10] J. Dulny, E. Kinnucan, J. Elliot, S. Mills, D. Farris, and J. Sullivan, "The artificial intelligence primer," 2018. Available: www.boozallen.com/s/insight/thought-leadership/the-artificial-intelligence-primer.html [Accessed June 15, 2023].

[11] "Medicine in the Digital Age," *Nature Medicine*, vol. 25, p. 1, Jan. 2019. https://doi.org/10.1038/s41591-018-0322-1

[12] M. De Graaf, "Will AI replace fertility doctors? Why computers are the only ones that can end the agony of failed IVF cycles, miscarriages, and risky multiple birth," 2018. Available: www.dailymail.co.uk/health/article-6257891/Study-finds-artificial-intelligence-better-doctor-crucial-stage-IVF.html [Accessed June 15, 2023].

[13] A. Avati, K. Jung, S. Harman, L. Downing, A. Ng, and N. H. Shah, "Improving palliative care with deep learning," *BMC Medical Informatics and Decision Making*, vol. 18, no. 4, pp. 55–64, 2018.

[14] B. Norgeot, B. S. Glicksberg, and A. J. Butte, "A call for deep-learning healthcare," *Nature Medicine*, vol. 25, no. 1, pp. 14–15, 2019.

[15] V. Gulshan, L. Peng, M. Coram, M. C. Stumpe, D. Wu, A. Narayanaswamy, S. Venugopalan, K. Widner, T. Madams, J. Cuadros, R. Kim, R. Raman, P. C. Nelson, J. L. Mega, and D. R. Webster, "Development and validation of a deep learning algorithm for detection of diabetic retinopathy in retinal fundus photographs," *JAMA*, vol. 316, pp. 2402–2410, Dec. 2016.

[16] J.-P. O. Li, H. Liu, D. S. Ting, S. Jeon, R. P. Chan, J. E. Kim, D. A. Sim, P. B. Thomas, H. Lin, Y. Chen, T. Sakomoto, A. Loewenstein, D. S. Lam, L. R. Pasquale, T. Y. Wong, L. A. Lam, and D. S. Ting, "Digital technology, tele-medicine and artificial intelligence in ophthalmology: A global perspective," *Progress in Retinal and Eye Research*, vol. 82 (2021): 100900. doi:10.1016/j.preteyeres.2020.100900.

[17] "The AI effect: How artificial intelligence is making health care more human," Oct. 2019. Available: https://mittrinsights.s3.amazonaws.com/ai-effect.pdf [Accessed June 15, 2023].

[18] "Geoff Hinton: On radiology." Available: https://youtu.be/2HMPRXstSvQ [Accessed June 26, 2023].

[19] K. Finley, "Robot radiologists will soon analyze your x-rays," Oct. 2015. Available: www.wired.com/2015/10/robot-radiologists-are-going-to-start-analyzing-x-rays/ [Accessed June 15, 2023].

[20] K. Chockley and E. Emanuel, "The end of radiology? Three threats to the future practice of radiology," *Journal of the American College of Radiology: JACR*, vol. 13, pp. 1415–1420, Dec. 2016.

[21] C. E. Kahn Jr, "We all need a little magic," *Radiology: Artificial Intelligence*, July 2019.

[22] RSNA, "Radiology: Artificial intelligence journal." Available: https://pubs.rsna.org/journal/ai [Accessed June 15, 2023].

[23] RSNA, "Radiology: Artificial intelligence blog." Available: https://pubs.rsna.org/page/ai/blog [Accessed: June 15, 2023].

[24] RSNA, "Introducing the artificial intelligence podcast." Available: https://pubs.rsna.org/page/ai/blog/2020/4/introducing_artificial_intelligence_podcast [Accessed June 15, 2023].

[25] "AI assistant blog — American college of radiology." Available: www.acrdsi.org/DSIBlog [Accessed July 20, 2023].

[26] Kaggle, "Kaggle." Available: www.kaggle.com/ [Accessed June 15, 2023].

[27] Kaggle, "RSNA-MICCAI brain tumor radiogenomic classification." Available: www.kaggle.com/competitions/rsna-miccai-brain-tumor-radiogenomic-classification [Accessed June 15, 2023].

[28] Kaggle, "RSNA intracranial hemorrhage detection." Available: www.kaggle.com/competitions/rsna-intracranial-hemorrhage-detection [Accessed June 15, 2023].

[29] Kaggle, "RSNA-STR pulmonary embolism detection." Available: www.kaggle.com/competitions/rsna-str-pulmonary-embolism-detection [Accessed June 15, 2023].

[30] Kaggle, "SIIM COVID-19 detection." Available: www.kaggle.com/competitions/siim-covid19-detection [Accessed June 15, 2023].

[31] Kaggle, "RSNA pneumonia detection challenge." Available: www.kaggle.com/competitions/rsna-pneumonia-detection-challenge [Accessed June 15, 2023].

[32] Kaggle, "SIIM-ISIC melanoma classification." Available: www.kaggle.com/competitions/siim-isic-melanoma-classification [Accessed June 15, 2023].

[33] RSNA, "RSNA pediatric bone age challenge 2017." Available: www.rsna.org/education/ai-resources-and-training/ai-image-challenge/RSNA-Pediatric-B-one-Age-Challenge-2017 [Accessed June 15, 2023].

[34] Kaggle, "SIIM-ACR pneumothorax segmentation." Available: www.kaggle.com/competitions/siim-acr-pneumothorax-segmentation [Accessed June 15, 2023].

[35] RSNA, "Cervical spine fractures AI detection challenge 2022." Available: www.rsna.org/education/ai-resources-and-training/ai-image-challenge/cervical-spine-fractures-ai-detection-challenge-2022 [Accessed June 15, 2023].

[36] RSNA, "Screening mammography breast cancer detection AI challenge." Available: www.rsna.org/education/ai-resources-and-training/ai-image-challenge/screening-mammography-breast-cancer-detection-ai-challenge [Accessed June 15, 2023].

[37] "Grand challenge." Available: https://grand-challenge.org/ [Accessed June 15, 2023].

[38] MICCAI, "MICCAI registered challenges." Available: www.miccai.org/special-interest-groups/challenges/miccai-registered-challenges/ [Accessed June 15, 2023].

[39] ImageCLEF, "ImageCLEF." Available: www.imageclef.org/ [Accessed June 15, 2023].

[40] RSNA, "Magicians' corner." Available: https://pubs.rsna.org/page/ai/magicians_corner [Accessed June 15, 2023].

[41] RSNA, "Magicians' corner GitHub repository." Available: https://github.com/RSNA/MagiciansCorner [Accessed June 15, 2023].

[42] M. J. Wood, N. A. Tenenholtz, J. R. Geis, M. H. Michalski, and K. P. Andriole, "The need for a machine learning curriculum for radiologists," *Journal of the American College of Radiology: JACR*, vol. 16, pp. 740–742, May 2019.

[43] "Choosing the right estimator – scikit-learn documentation." Available: https://scikit-learn.org/stable/tutorial/machine_learning_map/index.html [Accessed June 19, 2023].

[44] S. Leijnen and F. v. Veen, "The neural network zoo," *Proceedings*, vol. 47, p. 9, MDPI, 2020.

[45] Y. Bengio, I. J. Goodfellow, and A. Courville, "Deep learning," *Nature*, vol. 521, pp. 436–444, 2015.

[46] S. Hochreiter and J. Schmidhuber, "Long short-term memory," *Neural Computation*, vol. 9, pp. 1735–1780, Nov. 1997.

[47] I. Goodfellow, J. Pouget-Abadie, M. Mirza, B. Xu, D. Warde-Farley, S. Ozair, and Y. Bengio, "Generative adversarial networks, 1–9," *arXiv preprint arXiv:1406.2661*, 2014.

[48] A. Vaswani, N. Shazeer, N. Parmar, J. Uszkoreit, L. Jones, A. N. Gomez, L. Kaiser, and I. Polosukhin, "Attention is all you need," *Advances in Neural Information Processing Systems*, vol. 30, 2017.

[49] J. Devlin, M.-W. Chang, K. Lee, and K. Toutanova, "Bert: Pre-training of deep bidirectional transformers for language understanding," *arXiv preprint arXiv:1810.04805*, 2018.

[50] W. F. Wiggins and A. S. Tejani, "On the opportunities and risks of foundation models for natural language processing in radiology," *Radiology: Artificial Intelligence*, vol. 4, no. 4, p. e220119, 2022.

[51] J. Lee, W. Yoon, S. Kim, D. Kim, S. Kim, C. H. So, and J. Kang, "Biobert: A pre-trained biomedical language representation model for biomedical text mining," *Bioinformatics*, vol. 36, no. 4, pp. 1234–1240, 2020.

[52] J. M. Steinkamp, C. Chambers, D. Lalevic, H. M. Zafar, and T. S. Cook, "Toward complete structured information extraction from radiology reports using machine learning," *Journal of Digital Imaging*, vol. 32, pp. 554–564, 2019.

[53] A. Yan, J. McAuley, X. Lu, J. Du, E. Y. Chang, A. Gentili, and C.-N. Hsu, "Radbert: Adapting transformer-based language models to radiology," *Radiology: Artificial Intelligence*, vol. 4, no. 4, p. e210258, 2022.

[54] Z. Liu, A. Zhong, Y. Li, L. Yang, C. Ju, Z. Wu, C. Ma, P. Shu, C. Chen, S. Kim, *et al.*, "Radiology-GPT: A large language model for radiology," *arXiv preprint arXiv:2306.08666*, 2023.

[55] A. Dosovitskiy, L. Beyer, A. Kolesnikov, D. Weissenborn, X. Zhai, T. Unterthiner, M. Dehghani, M. Minderer, G. Heigold, S. Gelly, *et al.*, "An image is worth 16x16 words: Transformers for image recognition at scale," *arXiv preprint arXiv:2010.11929*, 2020.

[56] F. Zhuang, Z. Qi, K. Duan, D. Xi, Y. Zhu, H. Zhu, H. Xiong, and Q. He, "A comprehensive survey on transfer learning," *Proceedings of the IEEE*, vol. 109, no. 1, pp. 43–76, 2021.

[57] J. Deng, W. Dong, R. Socher, L.-J. Li, K. Li, and L. Fei-Fei, "ImageNet: A large-scale hierarchical image database," *CVPR09*, 2009. https://doi.org/10.1109/CVPR.2009.5206848

[58] F. Chollet, *Deep Learning with Python*, 2nd ed. Shelter Island: Manning Publications, 2021.

[59] A. Geron, *Hands-On Machine Learning with Scikit-Learn, Keras, and TensorFlow*, 3rd ed. Sebastopol, California: O'Reilly Media, Inc., 2022.

[60] A. P. Dhawan, *Medical Image Analysis*, 2nd ed. Hoboken, New Jersey: John Wiley & Sons, 2014.

[61] P. Suetens, *Fundamentals of Medical Imaging*, 3rd ed. Cambridge, England: Cambridge University Press, 2017.

[62] A. Krizhevsky, I. Sutskever, and G. E. Hinton, "Imagenet classification with deep convolutional neural networks," *Advances in Neural Information Processing Systems*, pp. 1097–1105, 2012.

[63] M. Xu, S. Yoon, A. Fuentes, and D. S. Park, "A comprehensive survey of image augmentation techniques for deep learning," *Pattern Recognition*, p. 109347, 2023.

[64] O. Ronneberger, P. Fischer, and T. Brox, "U-net: Convolutional networks for biomedical image segmentation," in *Medical Image Computing and Computer Assisted Intervention–MICCAI 2015: 18th International Conference, Munich, Germany, October 5-9, 2015, Proceedings, Part III 18*, pp. 234–241, Springer, 2015.

[65] V. Badrinarayanan, A. Kendall, and R. Cipolla, "Segnet: A deep convolutional encoder-decoder architecture for image segmentation," *IEEE Transactions on Pattern Analysis and Machine Intelligence*, vol. 39, no. 12, pp. 2481–2495, 2017.

[66] I. R. I. Haque and J. Neubert, "Deep learning approaches to biomedical image segmentation," *Informatics in Medicine Unlocked*, vol. 18, p. 100297, 2020.

[67] N. S. Punn and S. Agarwal, "Modality specific u-net variants for biomedical image segmentation: A survey," *Artificial Intelligence Review*, vol. 55, no. 7, pp. 5845–5889, 2022.

[68] M. H. Hesamian, W. Jia, X. He, and P. Kennedy, "Deep learning techniques for medical image segmentation: Achievements and challenges," *Journal of Digital Imaging*, vol. 32, pp. 582–596, 2019.

[69] N. Siddique, S. Paheding, C. P. Elkin, and V. Devabhaktuni, "U-net and its variants for medical image segmentation: A review of theory and applications," *IEEE Access*, vol. 9, pp. 82031–82057, 2021.

[70] A. Humeau-Heurtier, "Texture feature extraction methods: A survey," *IEEE Access*, vol. 7, pp. 8975–9000, 2019.

[71] K. Joshi and M. I. Patel, "Recent advances in local feature detector and descriptor: A literature survey," *International Journal of Multimedia Information Retrieval*, vol. 9, no. 4, pp. 231–247, 2020.

[72] P. D. Chang, T. T. Wong, and M. J. Rasiej, "Deep learning for detection of complete anterior cruciate ligament tear," *Journal of Digital Imaging*, vol. 32, no. 6, pp. 980– 986, 2019.

[73] Y. Mallya, J. Vijayananda, M. S. Vidya, K. Venugopal, and V. Mahajan, "Automatic delineation of anterior and posterior cruciate ligaments by combining deep learning and deformable atlas-based segmentation," in *Medical Imaging 2019: Biomedical Applications in Molecular, Structural, and Functional Imaging*, B. Gimi and A. Krol, Eds., vol. 10953, p. 1095321. International Society for Optics and Photonics. Bellingham, WA: SPIE, 2019.

[74] G. Litjens, T. Kooi, B. E. Bejnordi, A. A. A. Setio, F. Ciompi, M. Ghafoorian, J. A. van der Laak, B. van Ginneken, and C. I. Sánchez, "A survey on deep learning in medical image analysis," *Medical Image Analysis*, vol. 42, pp. 60–88, Dec. 2017.

[75] H. Hughes, M. O'Reilly, N. McVeigh, and R. Ryan, "The top 100 most cited articles on artificial intelligence in radiology: A bibliometric analysis," *Clinical Radiology*, vol. 78, no. 2, pp. 99–106, 2023.

[76] J. Egger, C. Gsaxner, A. Pepe, K. L. Pomykala, F. Jonske, M. Kurz, J. Li, and J. Kleesiek, "Medical deep learning—a systematic meta-review," *Computer Methods and Programs in Biomedicine*, vol. 221, p. 106874, 2022.

[77] "Awesome GAN for medical imaging." Available: https://github.com/xinario/awesome-gan-for-medical-imaging [Accessed June 16, 2023].

[78] L. X. Nguyen, P. Sone Aung, H. Q. Le, S.-B. Park, and C. S. Hong, "A new chapter for medical image generation: The stable diffusion method," in *2023 International Conference on Information Networking (ICOIN)*, Bangkok, Thailand, 2023, pp. 483–486, doi: 10.1109/ICOIN56518.2023.10049010.

[79] M. A. Azam, K. B. Khan, S. Salahuddin, E. Rehman, S. A. Khan, M. A. Khan, S. Kadry, and A. H. Gandomi, "A review on multimodal medical image fusion: Compendious analysis of medical modalities, multimodal databases, fusion techniques and quality metrics," *Computers in Biology and Medicine*, vol. 144, p. 105253, 2022.

[80] W. Tang, F. He, Y. Liu, and Y. Duan, "MATR: Multimodal medical image fusion via multiscale adaptive transformer," *IEEE Transactions on Image Processing*, vol. 31, pp. 5134–5149, 2022.

[81] R. J. Gillies, P. E. Kinahan, and H. Hricak, "Radiomics: Images are more than pictures, they are data," *Radiology*, vol. 278, no. 2, pp. 563–577, 2016.

[82] C. McCague, S. Ramlee, M. Reinius, I. Selby, D. Hulse, P. Piyatissa, V. Bura, M. Crispin-Ortuzar, E. Sala, and R. Woitek, "Introduction to radiomics for a clinical audience," *Clinical Radiology*, vol. 78, no. 2, pp. 83–98, 2023.

[83] M. A. Mazurowski, "Radiogenomics: What it is and why it is important," *Journal of the American College of Radiology*, vol. 12, no. 8, pp. 862–866, 2015.

[84] B. H. Van der Velden, H. J. Kuijf, K. G. Gilhuijs, and M. A. Viergever, "Explainable artificial intelligence (XAI) in deep learning-based medical image analysis," *Medical Image Analysis*, vol. 79, p. 102470, 2022.

[85] J. B. Kruskal, S. Berkowitz, J. R. Geis, W. Kim, P. Nagy, and K. Dreyer, "Big data and machine learning—strategies for driving this bus: A summary of the 2016 intersociety summer conference," *Journal of the American College of Radiology*, vol. 14, no. 6, pp. 811–817, 2017.

[86] "Introduction to data-centric AI." Available: https://dcai. csail.mit.edu/ [Accessed July 21, 2023].

[87] D. D. Hirsch, "The glass house effect: Big data, the new oil, and the power of analogy," *Maine Law Review*, vol. 66, p. 373, 2013.

[88] V. Sorin, Y. Barash, E. Konen, and E. Klang, "Deep learning for natural language processing in radiology—fundamentals and a systematic review," *Journal of the American College of Radiology*, vol. 17, no. 5, pp. 639–648, 2020.

[89] A. Hosny, C. Parmar, J. Quackenbush, L. H. Schwartz, and H. J. Aerts, "Artificial intelligence in radiology," *Nature Reviews Cancer*, vol. 18, no. 8, pp. 500–510, 2018.

[90] P. Mildenberger, M. Eichelberg, and E. Martin, "Introduction to the DICOM standard," *European Radiology*, vol. 12, pp. 920–927, 2002.

[91] R. H. Choplin, J. Boehme, and C. D. Maynard, "Picture archiving and communication systems: An overview," *Radiographics*, vol. 12, no. 1, pp. 127–129, 1992.

[92] C. Safran, M. Bloomrosen, W. E. Hammond, S. Labkoff, S. Markel-Fox, P. C. Tang, and D. E. Detmer, "Toward a national framework for the secondary use of health data: An American medical informatics association white paper," *Journal of the American Medical Informatics Association*, vol. 14, no. 1, pp. 1–9, 2007.

[93] A. Doan, A. Halevy, and Z. Ives, *Principles of Data Integration*. Amsterdam, NL: Elsevier, 2012.

[94] S. Candemir, X. V. Nguyen, L. R. Folio, and L. M. Prevedello, "Training strategies for radiology deep learning models in data-limited scenarios," *Radiology: Artificial Intelligence*, vol. 3, no. 6, p. e210014, 2021.

[95] M. L. Schiebler and C. Glide-Hurst, "Synthetic images are here to stay," *Radiology*, vol. 308, no. 1, p. e231098, 2023. doi:10.1148/radiol.231098.

[96] H. Jin, Y. Luo, P. Li, and J. Mathew, "A review of secure and privacy-preserving medical data sharing," *IEEE Access*, vol. 7, pp. 61656–61669, 2019.

[97] M. J. Sheller, B. Edwards, G. A. Reina, J. Martin, S. Pati, A. Kotrotsou, M. Milchenko, W. Xu, D. Marcus, R. R. Colen, *et al.*, "Federated learning in medicine: Facilitating multi-institutional collaborations without sharing patient data," *Scientific Reports*, vol. 10, no. 1, pp. 1–12, 2020.

[98] P. Rouzrokh, B. Khosravi, S. Faghani, M. Moassefi, D. V. Vera Garcia, Y. Singh, K. Zhang, G. M. Conte, and B. J. Erickson, "Mitigating bias in radiology machine learning: 1. Data handling," *Radiology: Artificial Intelligence*, vol. 4, no. 5, p. e210290, 2022.

[99] S. Chenthara, K. Ahmed, H. Wang, and F. Whittaker, "Security and privacy-preserving challenges of e-health solutions in cloud computing," *IEEE Access*, vol. 7, pp. 74361–74382, 2019.

[100] E. S. of Radiology (ESR), "The new EU general data protection regulation: What the radiologist should know," *Insights into Imaging*, vol. 8, pp. 295–299, 2017.

[101] W. D. Bidgood Jr, S. C. Horii, F. W. Prior, and D. E. Van Syckle, "Understanding and using DICOM, the data interchange standard for biomedical imaging," *Journal of the American Medical Informatics Association*, vol. 4, no. 3, pp. 199–212, 1997.

[102] "About health level seven international — hl7 international." Available: www.hl7.org/about/index.cfm [Accessed July 25, 2023].

[103] "Fhir v5.0.0." Available: https://hl7.org/fhir/ [Accessed July 25, 2023].

[104] M. Ayaz, M. F. Pasha, M. Y. Alzahrani, R. Budiarto, and D. Stiawan, "The fast health interoperability resources (FHIR) standard: Systematic literature review of implementations, applications, challenges and opportunities," *JMIR Medical Informatics*, vol. 9, no. 7, p. e21929, 2021.

[105] C. Park, S. C. You, H. Jeon, C. W. Jeong, J. W. Choi, and R. W. Park, "Development and validation of the radiology common data model (R-CDM) for the international standardization of medical imaging data," *Yonsei Medical Journal*, vol. 63, no. Suppl, p. S74, 2022.

[106] N. M. Safdar, J. D. Banja, and C. C. Meltzer, "Ethical considerations in artificial intelligence," *European Journal of Radiology*, vol. 122, p. 108768, 2020.

[107] K. Zhang, B. Khosravi, S. Vahdati, S. Faghani, F. Nugen, S. M. Rassoulinejad-Mousavi, M. Moassefi, J. M. M. Jagtap, Y. Singh, P. Rouzrokh, *et al.*, "Mitigating bias in radiology machine learning: 2. Model development," *Radiology: Artificial Intelligence*, vol. 4, no. 5, p. e220010, 2022.

[108] S. Faghani, B. Khosravi, K. Zhang, M. Moassefi, J. M. Jagtap, F. Nugen, S. Vahdati, S. P. Kuanar, S. M. Rassoulinejad-Mousavi, Y. Singh, *et al.*, "Mitigating bias in radiology machine learning: 3. Performance metrics," *Radiology: Artificial Intelligence*, vol. 4, no. 5, p. e220061, 2022.

[109] P. A. Keane and E. J. Topol, "With an eye to AI and autonomous diagnosis," *NPJ Digital Medicine*, vol. 1, no. 1, p. 40, 2018.

[110] M. Nagendran, Y. Chen, C. A. Lovejoy, A. C. Gordon, M. Komorowski, H. Harvey, E. J. Topol, J. P. Ioannidis, G. S. Collins, and M. Maruthappu, "Artificial intelligence versus clinicians: Systematic review of design, reporting standards, and claims of deep learning studies," *BMJ*, vol. 368, 2020.

[111] M. L. Richardson, E. R. Garwood, Y. Lee, M. D. Li, H. S. Lo, A. Nagaraju, X. V. Nguyen, L. Probyn, P. Rajiah, J. Sin, *et al.*, "Noninterpretive uses of artificial intelligence in radiology," *Academic Radiology*, vol. 28, no. 9, pp. 1225–1235, 2021.

[112] A. for Radiology: An Implementation Guide, "Products." Available: https://grand-challenge.org/aiforradiology/ [Accessed June 22, 2023].

[113] K. G. van Leeuwen, S. Schalekamp, M. J. Rutten, B. van Ginneken, and M. de Rooij, "Artificial intelligence in radiology: 100 commercially available products and their scientific evidence," *European Radiology*, vol. 31, pp. 3797–3804, 2021.

[114] A. DSI, "ACR data science institute AI central." Available: https://aicentral.acrdsi.org/ [Accessed June 22, 2023].

[115] US Food and Drug Administration (FDA), "CFR – Code of federal regulations title 21." Available: www.accessdata.fda. gov/scripts/cdrh/cfdocs/cfcfr/cfrsearch.cfm?fr=892.2050 [Accessed July 18, 2023].

[116] "Quantib® prostate – Quantib." Available: https://grand-challenge.org/aiforradiology/product/quantib-prostate/ [Accessed June 29, 2023].

[117] "Cina-ad – avicenna.ai." Available: https://grand-challenge.org/aiforradiology/product/avicenna-cina-chest-ad/ [Accessed June 29, 2023].

[118] "Livermultiscan – perspectum." Available: https://grand-challenge.org/aiforradiology/product/perspectum-diag-nostics-livermultiscan/ [Accessed June 29, 2023].

[119] "AI-rad companion prostate MR – siemens healthineers." Available: https://grand-challenge.org/aiforradiology/ product/siemens-rad-companion-prostate-mr/ [Accessed June 29, 2023].

[120] "Hepafat-scan – resonance health." Available: https:// grand-challenge.org/aiforradiology/product/resonance-health-hepafat-scan/ [Accessed June 29, 2023].

[121] "Intra-abdominal free gas (IFG) – Aidoc." Available: https://grand-challenge.org/aiforradiology/product/ aidoc-intra-abdominal-free-gas/ [Accessed June 29, 2023].

[122] "Lvivo toolbox – abdominal – DiA imaging analysis." Available: https://grand-challenge.org/aiforradiology/ product/dia-lvivo-abdominal/ [Accessed June 29, 2023].

[123] "Stonechecker – imaging biometrics." Available: https:// grand-challenge.org/aiforradiology/product/imaging-biometrics-stonechecker/ [Accessed June 29, 2023].

[124] "Mammoscreen® – therapixel." Available: https://grand-challenge.org/aiforradiology/product/therapixel-mam-moscreen/ [Accessed June 29, 2023].

[125] "Transpara – screenpoint medical." Available: https:// grand-challenge.org/aiforradiology/product/screenpoint-transpara/ [Accessed June 29, 2023].

[126] "Volparadensity – volpara health." Available: https:// grand-challenge.org/aiforradiology/product/volpara-health-volparadensity/ [Accessed June 29, 2023].

[127] "Koios DS – koios medical, inc." Available: https:// grand-challenge.org/aiforradiology/product/koios-ds/ [Accessed June 29, 2023].

[128] "Profound AI risk – ICAD." Available: https://grand-challenge.org/aiforradiology/product/icad-profound-ai-risk/ [Accessed June 29, 2023].

[129] "Lunit insight CXR – lunit." Available: https://grand-challenge.org/aiforradiology/product/lunit-insight-cxr3/ [Accessed June 29, 2023].

[130] "Qir-mr – casis – cardiac simulation & imaging software." Available: https://grand-challenge.org/aiforradiology/product/casis-qir-mr/ [Accessed June 29, 2023].

[131] "Heartflow FFRCT analysis – heartflow." Available: https://grand-challenge.org/aiforradiology/product/heartflow-ffrct-analysis/ [Accessed June 29, 2023].

[132] "Cardio AI – arterys." Available: https://grand-challenge.org/aiforradiology/product/arterys-cardio-ai/ [Accessed June 29, 2023].

[133] "Medis suite mr – medis medical imaging." Available: https://grand-challenge.org/aiforradiology/product/medis-suite-mr/ [Accessed June 29, 2023].

[134] "Cina-pe – avicenna.ai." Available: https://grand-challenge.org/aiforradiology/product/avicenna-cina-chest-pe/ [Accessed June 29, 2023].

[135] "Incidental pulmonary embolism (IPE) – Aidoc." Available: https://grand-challenge.org/aiforradiology/product/aidoc-incidental-pulmonary-embolism/ [Accessed June 29, 2023].

[136] "Annalise enterprise CXR – annalise.ai." Available: https://grand-challenge.org/aiforradiology/product/annalise-cxr/ [Accessed June 29, 2023].

[137] "Clearread xray – bone suppress – riverain technologies." Available: https://grand-challenge.org/aiforradiology/product/riverain-clearread-xray-bone-suppress/ [Accessed June 29, 2023].

[138] "Clearread xray – detect – riverain technologies." Available: https://grand-challenge.org/aiforradiology/product/riverain-clearread-xray-detect/ [Accessed June 29, 2023].

[139] "Clearread xray – confirm – riverain technologies." Available: https://grand-challenge.org/aiforradiology/product/riverain-clearread-xray-confirm/ [Accessed June 29, 2023].

[140] "Ib lab lama – imagebiopsy lab." Available: https://grand-challenge.org/aiforradiology/product/ib-lab-lama/ [Accessed June 29, 2023].

[141] "Rayvolve – azmed." Available: https://grand-challenge.org/aiforradiology/product/azmed-rayvolve/ [Accessed June 29, 2023].

[142] "Bonemri – mriguidance." Available: https://grand-challenge.org/aiforradiology/product/mriguidance-bonemri/ [Accessed June 29, 2023].

[143] "Boneview trauma – gleamer." Available: https://grand-challenge.org/aiforradiology/product/gleamer-ai-boneview-trauma/ [Accessed June 29, 2023].

[144] "C-spine (CSF) – Aidoc." Available: https://grand-challenge.org/aiforradiology/product/aidoc-c-spine/ [Accessed June 29, 2023].

[145] "Rib fractures (RIBFX) – Aidoc." Available: https://grand-challenge.org/aiforradiology/product/aidoc-rib-fractures/ [Accessed June 29, 2023].

[146] "Neurophet aqua – neurophet." Available: https://grand-challenge.org/aiforradiology/product/neurophet-aqua/ [Accessed June 29, 2023].

[147] "Intracranial hemorrhage (ICH) – Aidoc." Available: https://grand-challenge.org/aiforradiology/product/aidoc-ich/ [Accessed June 29, 2023].

[148] "Cina-ich – avicenna.ai." Available: https://grand-challenge.org/aiforradiology/product/avicenna-cina-ich/ [Accessed June 29, 2023].

[149] "Neuro AI – arterys." Available: https://grand-challenge.org/aiforradiology/product/arterys-neuro-ai/ [Accessed June 29, 2023].

[150] "Brain aneurysm (BA) – Aidoc." Available: https://grand-challenge.org/aiforradiology/product/aidoc-brain-aneurysm/ [Accessed June 29, 2023].

[151] "Koios ds – koios medical, inc." Available: https://grand-challenge.org/aiforradiology/product/koios-ds/ [Accessed June 29, 2023].

[152] "Columbo – smart soft healthcare." Available: https://grand-challenge.org/aiforradiology/product/smart-soft-columbo/ [Accessed June 29, 2023].

[153] "Pixelshine – algomedica." Available: https://grand-challenge.org/aiforradiology/product/algomedica-pixelshine/ [Accessed June 29, 2023].

[154] Y. Tadavarthi, V. Makeeva, W. Wagstaff, H. Zhan, A. Podlasek, N. Bhatia, M. Heilbrun, E. Krupinski, N. Safdar, I. Banerjee, *et al.*, "Overview of noninterpretive artificial intelligence models for safety, quality, workflow, and education applications in radiology practice," *Radiology: Artificial Intelligence*, vol. 4, no. 2, p. e210114, 2022.

[155] C. Mello-Thoms and C. A. Mello, "Clinical applications of artificial intelligence in radiology," *The British Journal of Radiology*, vol. 96, p. 20221031, 2023.

[156] J. Xu, E. Gong, J. Pauly, and G. Zaharchuk, "200x low-dose pet reconstruction using deep learning," *arXiv preprint arXiv:1712.04119*, 2017.

[157] C. Bermudez, A. J. Plassard, L. T. Davis, A. T. Newton, S. M. Resnick, and B. A. Landman, "Learning implicit brain MRI manifolds with deep learning," *Medical Imaging 2018: Image Processing*, vol. 10574, pp. 408–414, SPIE, 2018.

[158] D. Tamada, "Noise and artifact reduction for MRI using deep learning," *arXiv preprint arXiv:2002.12889*, 2020.

[159] T. Higaki, Y. Nakamura, F. Tatsugami, T. Nakaura, and K. Awai, "Improvement of image quality at CT and MRI using deep learning," *Japanese Journal of Radiology*, vol. 37, pp. 73–80, 2019.

[160] J. V. Manjon and P. Coupe, "MRI denoising using deep learning and non-local averaging," *arXiv preprint arXiv: 1911.04798*, 2019.

[161] Y. Wang, B. Yu, L. Wang, C. Zu, D. S. Lalush, W. Lin, X. Wu, J. Zhou, D. Shen, and L. Zhou, "3d conditional generative adversarial networks for high-quality pet image estimation at low dose," *Neuroimage*, vol. 174, pp. 550–562, 2018.

[162] J. Ouyang, K. T. Chen, E. Gong, J. Pauly, and G. Zaharchuk, "Ultra-low-dose pet reconstruction using generative adversarial network with feature matching and taskspecific perceptual loss," *Medical Physics*, vol. 46, no. 8, pp. 3555–3564, 2019.

[163] M. Eberhard and H. Alkadhi, "Machine learning and deep neural networks: Applications in patient and scan preparation, contrast medium, and radiation dose optimization," *Journal of Thoracic Imaging*, vol. 35, pp. S17–S20, 2020.

[164] N. Saltybaeva, B. Schmidt, A. Wimmer, T. Flohr, and H. Alkadhi, "Precise and automatic patient positioning in computed tomography: Avatar modeling of the patient surface using a 3-dimensional camera," *Investigative Radiology*, vol. 53, no. 11, pp. 641–646, 2018.

[165] R. Booij, M. van Straten, A. Wimmer, and R. P. Budde, "Automated patient positioning in CT using a 3D camera for body contour detection: Accuracy in pediatric patients," *European Radiology*, vol. 31, pp. 131–138, 2021.

[166] R. Booij, R. P. Budde, M. L. Dijkshoorn, and M. van Straten, "Accuracy of automated patient positioning in CT using a 3D camera for body contour detection," *European Radiology*, vol. 29, pp. 2079–2088, 2019.

[167] P. Hejduk, R. Sexauer, C. Ruppert, K. Borkowski, J. Unkelbach, and N. Schmidt, "Automatic and standardized quality assurance of digital mammography and tomosynthesis with deep convolutional neural networks," *Insights into Imaging*, vol. 14, no. 1, pp. 1–9, 2023.

[168] "Intellimammo™ – densitas." Available: https://grand-challenge.org/aiforradiology/product/densitas-intelli-mammo/ [Accessed July 19, 2023].

[169] "Mammography quality standards act and program – FDA." Available: www.fda.gov/radiation-emitting-prod-ucts/mammography-quality-standards-act-and-program [Accessed June 29, 2023].

[170] "EUREF – European reference organisation for quality assured breast screening and diagnostic services." Available: https://euref.org/#:~:text=EUREF%2C%20the%20European%20Reference%20Organisation,basis%20for%20over%2030%20years [Accessed July 19, 2023].

[171] "02hecs267-eng.pdf." Available: www.canada.ca/content/dam/hc-sc/migration/hc-sc/ewh-semt/alt_formats/hecs-sesc/pdf/pubs/radiation/02hecs-sesc267/02hecs267-eng.pdf [Accessed July 19, 2023].

[172] N. A. T. Wadden and C. Hapgood, "Canadian association of radiologists mammography accreditation program—clinical image assessment," *Canadian Association of Radiologists Journal*, vol. 73, no. 1, pp. 157–163, 2022.

[173] L. M. Prevedello, B. S. Erdal, J. L. Ryu, K. J. Little, M. Demirer, S. Qian, and R. D. White, "Automated critical test findings identification and online notification system using artificial intelligence in imaging," *Radiology*, vol. 285, no. 3, pp. 923–931, 2017.

[174] S. Goldberg-Stein and V. Chernyak, "Adding value in radiology reporting," *Journal of the American College of Radiology*, vol. 16, no. 9, pp. 1292–1298, 2019.

[175] T. Cai, A. A. Giannopoulos, S. Yu, T. Kelil, B. Ripley, K. K. Kumamaru, F. J. Rybicki, and D. Mitsouras, "Natural language processing technologies in radiology research and clinical applications," *Radiographics*, vol. 36, no. 1, pp. 176–191, 2016.

[176] N. Mehta, V. Harish, K. Bilimoria, F. Morgado, S. Ginsburg, M. Law, and S. Das, "Knowledge of and attitudes on artificial intelligence in healthcare: A provincial survey study of medical students [version 1]," *MedEdPublish*, vol. 10, p. 75, 2021. https://doi.org/10.15694/mep.2021.000075.1.

[177] S. M. Santomartino and H. Y. Paul, "Systematic review of radiologist and medical student attitudes on the role and impact of AI in radiology," *Academic Radiology*, S1076–6332(21)00624-3. 29 Jan. 2022. doi:10.1016/j.acra.2021.12.032.

[178] E. M. Weisberg and E. K. Fishman, "Developing a curriculum in artificial intelligence for emergency radiology," *Emergency Radiology*, vol. 27, no. 4, pp. 359–360, 2020.

[179] M. Huisman, E. Ranschaert, W. Parker, D. Mastrodicasa, M. Koci, D. Pinto de Santos, F. Coppola, S. Morozov, M. Zins, C. Bohyn, *et al.*, "An international survey on AI in radiology in 1041 radiologists and radiology residents part 2: Expectations, hurdles to implementation, and education," *European Radiology*, vol. 31, no. 11, pp. 8797–8806, 2021.

[180] A. S. Tejani, H. Elhalawani, L. Moy, M. Kohli, and C. E. Kahn Jr, "Artificial intelligence and radiology education," *Radiology: Artificial Intelligence*, vol. 5, no. 1, p. e220084, 2022.

[181] A. L. Lindqwister, S. Hassanpour, J. Levy, and J. M. Sin, "AI-rads: Successes and challenges of a novel artificial intelligence curriculum for radiologists across different delivery formats," *Frontiers in Medical Technology*, vol. 4, p. 1007708, 2023.

[182] A. C. of Radiology, "Define-AI use case directory." Available: www.acrdsi.org/DSI-Services/Define-AI [Accessed June 28, 2023].

[183] B. Ngo, D. Nguyen, and E. vanSonnenberg, "The cases for and against artificial intelligence in the medical school curriculum," *Radiology: Artificial Intelligence*, vol. 4, no. 5, p. e220074, 2022.

[184] V. Gowda, S. G. Jordan, and O. A. Awan, "Artificial intelligence in radiology education: A longitudinal approach," *Academic Radiology*, vol. 29, no. 5, pp. 788–790, 2022.

[185] J. Perchik and S. Tridandapani, "AI/ML education in radiology: Accessibility is key," *Academic Radiology*, vol. 30, no. 7, pp. 1491–1492, 2023.

[186] N. V. Salastekar, C. Maxfield, T. N. Hanna, E. A. Krupinski, D. Heitkamp, and L. J. Grimm, "Artificial intelligence/machine learning education in radiology: Multiinstitutional

survey of radiology residents in the United States," *Academic Radiology*, vol. 30, no. 7, pp. 1481–1487, 2023. doi:10.1016/j.acra.2023.01.005.

[187] S. P. Garin, V. Zhang, J. Jeudy, V. S. Parekh, and H. Y. Paul, "Systematic review of radiology residency AI curricula: Preparing future radiologists for the AI era," *Journal of the American College of Radiology*, vol. 20, no. 6, pp. 561–569, 2023.

[188] M. D. Li and B. P. Little, "Appropriate reliance on AI in radiology education," *Journal of the American College of Radiology*, S1546-1440(23)00422-2, 29 Jun. 2023. doi:10.1016/j.jacr.2023.04.019

[189] W. F. Wiggins, M. T. Caton, K. Magudia, S. A. Glomski, E. George, M. H. Rosenthal, G. C. Gaviola, and K. P. Andriole, "Preparing radiologists to lead in the era of artificial intelligence: Designing and implementing a focused data science pathway for senior radiology residents," *Radiology: Artificial Intelligence*, vol. 2, Nov 2020.

[190] W. F. Wiggins, "William F. Wiggins." Available: https://wfwiggins.github.io/ [Accessed June 15, 2023].

[191] O. Marques, "Deep learning workflow: Tips, tricks, and often forgotten steps." Available: www.kdnuggets.com/2020/09/mathworks-deep-learning-workflow.html [Accessed June 28, 2023].

[192] H. Rosling, O. Rosling, and A. Rönnlund, *Factfulness: Ten Reasons We're Wrong About the World—And Why Things Are Better Than You Think*. New York, NY: Flatiron Books, 2018.

[193] E. Breck, S. Cai, E. Nielsen, M. Salib, and D. Sculley, "The ML test score: A rubric for ML production readiness and technical debt reduction," in *2017 IEEE International Conference on Big Data (Big Data)* (pp. 1123–1132). IEEE, December 2017.

[194] Google, "Production ML systems." Available: https://developers.google.com/machine-learning/crash-course/production-ml-systems [Accessed June 15, 2023].

[195] P. Lakhani, D. L. Gray, C. R. Pett, P. Nagy, and G. Shih, "Hello world deep learning in medical imaging," *Journal of Digital Imaging*, vol. 31, no. 3, pp. 283–289, 2018.

[196] "Hello world for deep learning." Available: www.kaggle.com/code/wfwiggins203/hello-world-for-deep-learning-siim/notebook [Accessed June 15, 2023].

[197] RSNA, "MedNISTClassify." Available: https://colab. research.google.com/github/RSNA/MagiciansCorner/ blob/master/MedNISTClassify.ipynb [Accessed June 15, 2023].

[198] O. Marques and A. Corbin, "Deep learning for medical imaging: A recipe for success." 2022. Available: https://cdn. ymaws.com/siim.org/resource/resmgr/siim2022/docu- ments/abstract/poster/apppost-_marques,_oge__deep_. pdf.

[199] M. J. Cardoso, W. Li, R. Brown, N. Ma, E. Kerfoot, Y. Wang, B. Murrey, A. Myronenko, C. Zhao, D. Yang, V. Nath, Y. He, Z. Xu, A. Hatamizadeh, A. Myronenko, W. Zhu, Y. Liu, M. Zheng, Y. Tang, I. Yang, M. Zephyr, B. Hashemian, S. Alle, M. Z. Darestani, C. Budd, M. Modat, T. Vercauteren, G. Wang, Y. Li, Y. Hu, Y. Fu, B. Gorman, H. Johnson, B. Genereaux, B. S. Erdal, V. Gupta, A. Diaz-Pinto, A. Dourson, L. Maier-Hein, P. F. Jaeger, M. Baumgartner, J. Kalpathy- Cramer, M. Flores, J. Kirby, L. A. D. Cooper, H. R. Roth, D. Xu, D. Bericat, R. Floca, S. K. Zhou, H. Shuaib, K. Farahani, K. H. Maier-Hein, S. Aylward, P. Dogra, S. Ourselin, and A. Feng, "Monai: An open-source framework for deep learn- ing in healthcare," *arXiv Preprint arXiv:2211.02701*, 2022.

[200] L. Maier-Hein, M. Eisenmann, A. Reinke, S. Onogur, M. Stankovic, P. Scholz, T. Arbel, H. Bogunovic, A. P. Bradley, A. Carass, *et al.*, "Why rankings of biomedical image analy- sis competitions should be interpreted with care," *Nature Communications*, vol. 9, no. 1, p. 5217, 2018.

[201] RSNA, "Checklist for artificial intelligence in medical imaging (CLAIM)." Available: https://pubs.rsna.org/ page/ai/claim [Accessed June 20, 2023].

[202] P. M. Bossuyt, J. B. Reitsma, D. E. Bruns, C. A. Gatsonis, P. P. Glasziou, L. Irwig, J. G. Lijmer, D. Moher, D. Rennie, H. C. de Vet, H. Y. Kressel, N. Rifai, R. M. Golub, D. G. Altman, L. Hooft, D. A. Korevaar, and J. F. A. Cohen, "Stard 2015: An updated list of essential items for reporting diagnostic accu- racy studies," *Radiology*, vol. 277, no. 3, pp. 826–832, 2015.

[203] D. A. Bluemke, L. Moy, M. A. Bredella, B. B. Ertl-Wagner, K. J. Fowler, V. J. Goh, E. F. Halpern, C. P. Hess, M. L. Schiebler, and C. R. Weiss, "Assessing radiology research on artificial intelligence: A brief guide for authors, review- ers, and readers—from the radiology editorial board," *Radiology*, vol. 294, no. 3, pp. 487–489, 2020.

[204] A. J. Thirunavukarasu, D. S. J. Ting, K. Elangovan, L. Gutierrez, T. F. Tan, and D. S. W. Ting, "Large language models in medicine," *Nature Medicine*, pp. 1–11, 2023.

[205] S. Bird, M. Dudík, R. Edgar, B. Horn, R. Lutz, V. Milan, M. Sameki, H. Wallach, and K. Walker, "Fairlearn: A toolkit for assessing and improving fairness in AI," *Microsoft, Technical Report MSR-TR-2020-32*, N/A, 2020.

[206] R. K. Bellamy, K. Dey, M. Hind, S. C. Hoffman, S. Houde, K. Kannan, P. Lohia, J. Martino, S. Mehta, A. Mojsilović, *et al.*, "AI fairness 360: An extensible toolkit for detecting and mitigating algorithmic bias," *IBM Journal of Research and Development*, vol. 63, no. 4/5, pp. 4–1, 2019.

[207] C. Shah, D. Nachand, C. Wald, and P.-H. Chen, "Keeping patient data secure in the age of radiology AI: Cybersecurity considerations and future directions," *Journal of the American College of Radiology*, N/A, 2023.

[208] FDA, "De Novo classification request." Available: www.fda.gov/medical-devices/premarket-submissions-selecting-and-preparing-correct-submission/de-novo-classification-request [Accessed June 21, 2023].

[209] E. Commission, "Regulation (EU) 2017/745 of the European Parliament and of the council of 5 April 2017 on medical devices." Available: https://eur-lex.europa.eu/eli/reg/2017/745/2020-04-24 [Accessed June 21, 2023].

[210] E. Commission, "Regulation (EU) 2017/745 of the European Parliament and of the council of 5 April 2017 on in vitro diagnostic devices." Available: https://eur-lex.europa.eu/legal-content/EN/TXT/PDF/?uri=CELEX:32017R0746 [Accessed June 21, 2023].

[211] E. Commission, "Proposal for a regulation of the European Parliament and of the council laying down harmonised rules on artificial intelligence (artificial intelligence act) and Amending certain union legislative acts." Available: https://eur-lex.europa.eu/legal-content/EN/TXT/?uri=CELEX:52021PC0206 [Accessed June 21, 2023]

[212] F. Cabitza, R. Rasoini, and G. F. Gensini, "Unintended consequences of machine learning in medicine," *JAMA*, vol. 318, no. 6, pp. 517–518, 2017.

[213] F. C. Kitamura and O. Marques, "Trustworthiness of artificial intelligence models in radiology and the role of explainability," *Journal of the American College of Radiology*, vol. 18, no. 8, pp. 1160–1162, 2021.

[214] Z. Xue, F. Yang, S. Rajaraman, G. Zamzmi, and S. Antani, "Cross dataset analysis of domain shift in CXR lung region detection," *Diagnostics*, vol. 13, no. 6, p. 1068, 2023.

[215] R. Lacson, M. Eskian, A. Licaros, N. Kapoor, and R. Khorasani, "Machine learning model drift: Predicting diagnostic imaging follow-up as a case example," *Journal of the American College of Radiology*, vol. 19, no. 10, pp. 1162–1169, 2022.

[216] Z. Lv and F. Piccialli, "The security of medical data on internet based on differential privacy technology," *ACM Transactions on Internet Technology*, vol. 21, no. 3, pp. 1–18, 2021.

[217] "MONAI model zoo." Available: https://monai.io/model-zoo.html [Accessed July 25, 2023].

[218] M. Wu, X. Zhang, J. Ding, H. Nguyen, R. Yu, M. Pan, and S. T. Wong, "Evaluation of inference attack models for deep learning on medical data," *arXiv preprint arXiv:2011.00177*, 2020.

[219] J. R. Geis, A. P. Brady, C. C. Wu, J. Spencer, E. Ranschaert, J. L. Jaremko, S. G. Langer, A. B. Kitts, J. Birch, W. F. Shields, *et al.*, "Ethics of artificial intelligence in radiology: Summary of the joint European and North American multisociety statement," *Canadian Association of Radiologists Journal*, vol. 70, no. 4, pp. 329–334, 2019.

[220] K. Badal, C. M. Lee, and L. J. Esserman, "Guiding principles for the responsible development of artificial intelligence tools for healthcare," *Communications Medicine*, vol. 3, no. 1, p. 47, 2023.

[221] T. Davenport and R. Kalakota, "The potential for artificial intelligence in healthcare," *Future Healthcare Journal*, vol. 6, no. 2, p. 94, 2019.

[222] D. S. Char, N. H. Shah, and D. Magnus, "Implementing machine learning in health care—addressing ethical challenges," *The New England Journal of Medicine*, vol. 378, no. 11, p. 981, 2018.

[223] J. Grunhut, A. T. Wyatt, and O. Marques, "Educating future physicians in artificial intelligence (AI): An integrative review and proposed changes," *Journal of Medical Education and Curricular Development*, vol. 8, p. 23821205211036836, 2021.

[224] J. Krive, M. Isola, L. Chang, T. Patel, M. Anderson, and R. Sreedhar, "Grounded in reality: Artificial intelligence in medical education," *JAMIA Open*, vol. 6, no. 2, p. ooad037, 2023.

[225] A. Lecler, L. Duron, and P. Soyer, "Revolutionizing radiology with GPT-based models: Current applications, future possibilities and limitations of ChatGPT," *Diagnostic and Interventional Imaging*, vol. 104, no. 6, pp. 269–274, 2023.

[226] K. M. Boehm, P. Khosravi, R. Vanguri, J. Gao, and S. P. Shah, "Harnessing multimodal data integration to advance precision oncology," *Nature Reviews Cancer*, vol. 22, no. 2, pp. 114–126, 2022.

[227] L. R. Soenksen, Y. Ma, C. Zeng, L. Boussioux, K. Villalobos Carballo, L. Na, H. M. Wiberg, M. L. Li, I. Fuentes, and D. Bertsimas, "Integrated multimodal artificial intelligence framework for healthcare applications," *NPJ Digital Medicine*, vol. 5, no. 1, p. 149, 2022.

[228] S.-C. Huang, A. Pareek, S. Seyyedi, I. Banerjee, and M. P. Lungren, "Fusion of medical imaging and electronic health records using deep learning: A systematic review and implementation guidelines," *NPJ Digital Medicine*, vol. 3, no. 1, p. 136, 2020.

[229] J. Li, A. Dada, J. Kleesiek, and J. Egger, "ChatGPT in healthcare: A taxonomy and systematic review," *medRxiv*, 2023.03.30.23287899, 2023. https://doi.org/10.1101/2023.03.30.23287899

[230] P. Rajpurkar and M. P. Lungren, "The current and future state of AI interpretation of medical images," *New England Journal of Medicine*, vol. 388, no. 21, pp. 1981–1990, 2023.

[231] M. Moor, O. Banerjee, Z. S. H. Abad, H. M. Krumholz, J. Leskovec, E. J. Topol, and P. Rajpurkar, "Foundation models for generalist medical artificial intelligence," *Nature*, vol. 616, no. 7956, pp. 259–265, 2023.

[232] R. Challen, J. Denny, M. Pitt, L. Gompels, T. Edwards, and K. Tsaneva-Atanasova, "Artificial intelligence, bias and clinical safety," *BMJ Quality & Safety*, vol. 28, no. 3, pp. 231–237, 2019.

[233] E. Topol, *Deep Medicine: How Artificial Intelligence Can Make Healthcare Human Again*. New York, NY: Hachette, 2019.

[234] K. L. Holley and S. Becker, *AI-first Healthcare*. O'Reilly Media, Inc., 2021.

[235] P. Rajpurkar, E. Chen, O. Banerjee, and E. J. Topol, "AI in health and medicine," *Nature Medicine*, vol. 28, no. 1, pp. 31–38, 2022.

[236] M. Mitchell, "Why AI is harder than we think." 2021. Available: https://arxiv.org/abs/2104.12871.

[237] B. Meskó and M. Görög, "A short guide for medical professionals in the era of artificial intelligence," *NPJ Digital Medicine*, vol. 3, no. 1, 2020.

[238] A. Esteva, A. Robicquet, B. Ramsundar, V. Kuleshov, M. DePristo, K. Chou, C. Cui, G. Corrado, S. Thrun, and J. Dean, "A guide to deep learning in healthcare," *Nature Medicine*, vol. 25, no. 1, pp. 24–29, 2019.

[239] S. Mukherjee, "A.I. versus M.D." Mar. 2017 Available: www.newyorker.com/magazine/2017/04/03/ai-versus-md.

[240] A. Ng, "AI for everyone." Available: www.deeplearning.ai/courses/ai-for-everyone/ [Accessed June 26, 2023].

[241] J. Banja, "AI hype and radiology: A plea for realism and accuracy," *Radiology: Artificial Intelligence*, vol. 2, no. 4, 2020.

[242] F. Pesapane, M. Codari, and F. Sardanelli, "Artificial intelligence in medical imaging: Threat or opportunity? Radiologists again at the forefront of innovation in medicine," *European Radiology Experimental*, vol. 2, pp. 1–10, 2018.

[243] J. Brownlee, "Machine learning books." Available: https://machinelearningmastery.com/machine-learning-books/ [Accessed June 26, 2023].

[244] J. Misiti, "Awesome machine learning." Available: https://github.com/josephmisiti/awesome-machine-learning/ [Accessed June 26, 2023].

[245] A. Ng, *Machine Learning Yearning: Technical Strategy for AI Engineers in the Era of Deep Learning.* N/A, 2019.

[246] M. Nielsen, "Neural networks and deep learning." Available: http://neuralnetworksanddeeplearning.com/.

[247] I. Goodfellow, Y. Bengio, and A. Courville, *Deep Learning.* Cambridge, Massachusetts: MIT Press, 2016.

[248] Google, "Google machine learning education." Available: https://developers.google.com/machine-learning [Accessed June 27, 2023].

[249] C. Kozyrkov, "Making friends with machine learning." Available: https://youtu.be/1vkb7BCMQd0 [Accessed June 27, 2023].

[250] 3Blue1Brown, "Neural networks." Available: www.youtube.com/playlist?list=PLZHQObOWTQDNU6R1_67000Dx_ZCJB-3pi [Accessed June 25, 2023].

[251] "CS231n convolutional neural networks for visual recognition." Available: https://cs231n.github.io/convolutional-networks/ [Accessed June 27, 2023].

[252] "Understanding convolutions – colah's blog." Available: http://colah.github.io/posts/2014-07-Understanding-Convolutions/ [Accessed June 22, 2023].

[253] "Understanding LSTM networks – colah's blog." Available: https://colah.github.io/posts/2015-08-Understanding-LSTMs/ [Accessed June 22, 2023].

[254] W. Birkfellner, *Applied Medical Image Processing: A Basic Course*, 2nd ed. Boca Raton, FL: CRC Press, an imprint of Taylor and Francis, 2014.

[255] K. M. Martensen, *Radiographic Image Analysis*. Amsterdam, NL: Elsevier Health Sciences, 2013.

[256] J. Mandell, *Core Radiology*. Cambridge, England: Cambridge University Press, 2013.

[257] E. R. Ranschaert, S. Morozov, and P. R. Algra, *Artificial Intelligence in Medical Imaging: Opportunities, Applications and Risks*. Heidelberg, Germany: Springer, 2019.

[258] A. Esteva, K. Chou, S. Yeung, N. Naik, A. Madani, A. Mottaghi, Y. Liu, E. Topol, J. Dean, and R. Socher, "Deep learning-enabled medical computer vision," *NPJ Digital Medicine*, vol. 4, no. 1, p. 5, 2021.

[259] "ImageJ." Available: https://imagej.net/ij/ [Accessed June 22, 2023].

[260] "Fiji." Available: https://imagej.net/software/fiji/ [Accessed June 22, 2023].

[261] "MATLAB – mathworks." Available: www.mathworks.com/products/matlab.html [Accessed June 22, 2023].

[262] A. Kaehler and G. Bradski, *Learning OpenCV 3: Computer Vision in C++ with the OpenCV Library*. Sebastopol, California: O'Reilly Media, Inc., 2017.

[263] "Pillow (PIL fork) documentation." Available: https://pillow.readthedocs.io/en/stable/ [Accessed June 29, 2023].

[264] A. Nguyen, *Hands-On Healthcare Data*. Sebastopol, California: O'Reilly Media, Inc., 2022.

[265] O. Diaz, K. Kushibar, R. Osuala, A. Linardos, L. Garrucho, L. Igual, P. Radeva, F. Prior, P. Gkontra, and K. Lekadir, "Data preparation for artificial intelligence in medical imaging: A comprehensive guide to open-access platforms and tools," *Physica Medica*, vol. 83, pp. 25–37, 2021.

[266] M. J. Willemink, W. A. Koszek, C. Hardell, J. Wu, D. Fleischmann, H. Harvey, L. R. Folio, R. M. Summers, D. L. Rubin, and M. P. Lungren, "Preparing medical imaging data for machine learning," *Radiology*, vol. 295, no. 1, pp. 4–15, 2020.

[267] "Monai label end-to-end tutorial series." Available: https://github.com/Project-MONAI/tutorials/tree/main/monai-label [Accessed June 27, 2023].

[268] A. Diaz-Pinto, S. Alle, V. Nath, Y. Tang, A. Ihsani, M. Asad, F. Pérez-García, P. Mehta, W. Li, M. Flores, H. R. Roth, T. Vercauteren, D. Xu, P. Dogra, S. Ourselin, A. Feng, and M. J. Cardoso, "Monai label: A framework for ai-assisted interactive labeling of 3D medical images," *arXiv preprint arXiv*, 2203.12362. 2022 Mar 2023.

[269] C. P. Langlotz, B. Allen, B. J. Erickson, J. Kalpathy-Cramer, K. Bigelow, T. S. Cook, A. E. Flanders, M. P. Lungren, D. S. Mendelson, J. D. Rudie, *et al.*, "A roadmap for foundational research on artificial intelligence in medical imaging: From the 2018 NIH/RSNA/ACR/the academy workshop," *Radiology*, vol. 291, no. 3, pp. 781–791, 2019.

[270] B. Allen Jr, S. E. Seltzer, C. P. Langlotz, K. P. Dreyer, R. M. Summers, N. Petrick, D. Marinac-Dabic, M. Cruz, T. K. Alkasab, R. J. Hanisch, *et al.*, "A road map for translational research on artificial intelligence in medical imaging: From the 2018 national institutes of health/RSNA/ACR/the academy workshop," *Journal of the American College of Radiology*, vol. 16, no. 9, pp. 1179–1189, 2019.

[271] M. Khunte, A. Chae, R. Wang, R. Jain, Y. Sun, J. Sollee, Z. Jiao, and H. Bai, "Trends in clinical validation and usage of US Food and Drug Administration-cleared artificial intelligence algorithms for medical imaging," *Clinical Radiology*, vol. 78, no. 2, pp. 123–129, 2023.

[272] M. Milam and C. Koo, "The current status and future of FDA-approved artificial intelligence tools in chest radiology in the United States," *Clinical Radiology*, vol. 78, no. 2, pp. 115–122, 2023.

[273] D. P. Stonko and C. W. Hicks, "Mature AI/ML-enabled medical tools impacting vascular surgical care: A scoping review of late-stage, FDA approved/cleared technologies relevant to vascular surgeons," in *Seminars in Vascular Surgery*. Amsterdam, NL: Elsevier, 2023.

[274] FDA, "Artificial intelligence and machine learning in software as a medical device." Available: www.fda.gov/medical-devices/software-medical-device-samd/artificial-intelligence-and-machine-learning-software-medical-device [Accessed June 28, 2023].

[275] SIIM, "SIIM hackathon." Available: https://siim.org/page/siim_hackathon [Accessed June 15, 2023].

[276] RSNA, "RSNA education catalog." Available: https://education.rsna.org/diweb/catalog [Accessed June 15, 2023].

[277] ACR, "AI-LAB." Available: https://ailab.acr.org/ [Accessed June 15, 2023].

[278] SIIM, "National imaging informatics curriculum." Available: https://siim.org/page/niic [Accessed June 15, 2023].

[279] "Fundamentals of machine learning for healthcare – Coursera." Available: www.coursera.org/learn/fundamental-machine-learning-healthcare [Accessed July 19, 2023].

[280] RSNA, "RSNA AI certificate program." Available: www.rsna.org/AI-certificate. [Accessed June 15, 2023].

[281] J. Howard and S. Gugger, *Deep Learning for Coders with Fastai and PyTorch*. Sebastopol, California: O'Reilly Media, Inc., 2020.

[282] Kaggle, "Python." Available: www.kaggle.com/learn/python [Accessed June 28, 2023].

[283] Kaggle, "Intro to machine learning." Available: www.kaggle.com/learn/intro-to-machine-learning [Accessed June 28, 2023].

[284] Kaggle, "Intermediate machine learning." Available: www.kaggle.com/learn/intermediate-machine-learning [Accessed June 28, 2023].

[285] Kaggle, "Intro to deep learning." Available: www.kaggle.com/learn/intro-to-deep-learning [Accessed June 28, 2023].

[286] "Awesome python." Available: https://awesome-python.com/ [Accessed June 28, 2023].

[287] W3Schools, "Git tutorial." Available: www.w3schools.com/git/ [Accessed June 28, 2023].

[288] "PyTorch tutorials." Available: https://pytorch.org/tutorials/ [Accessed June 29, 2023].

[289] "Introduction to TensorFlow." Available: www.tensorflow.org/learn [Accessed June 26, 2023].

[290] "Python for nonprogrammers." Available: https://wiki.python.org/moin/BeginnersGuide/NonProgrammers [Accessed June 28, 2023].

[291] T. Boeken, J. Feydy, A. Lecler, P. Soyer, A. Feydy, M. Barat, and L. Duron, "Artificial intelligence in diagnostic and interventional radiology: Where are we now?" *Diagnostic and Interventional Imaging*, vol. 104, no. 1, pp. 1–5, January 2023.

[292] C. Garbin and O. Marques, "Assessing methods and tools to improve reporting, increase transparency, and reduce failures in machine learning applications in health care," *Radiology: Artificial Intelligence*, vol. 4, no. 2, p. e210127, 2022.

[293] E. Wu, K. Wu, R. Daneshjou, D. Ouyang, D. E. Ho, and J. Zou, "How medical AI devices are evaluated: Limitations and recommendations from an analysis of FDA approvals," *Nature Medicine*, vol. 27, no. 4, pp. 582–584, 2021.

Index

Note: Page numbers in *italics* indicate a figure and page numbers in **bold** indicate a table on the corresponding page.